BEGINNERS'

GUIDE TO WARGAMING

BRUCE QUARRIE

Great battles on land, at sea and in the air!
No previous playing experience needed!

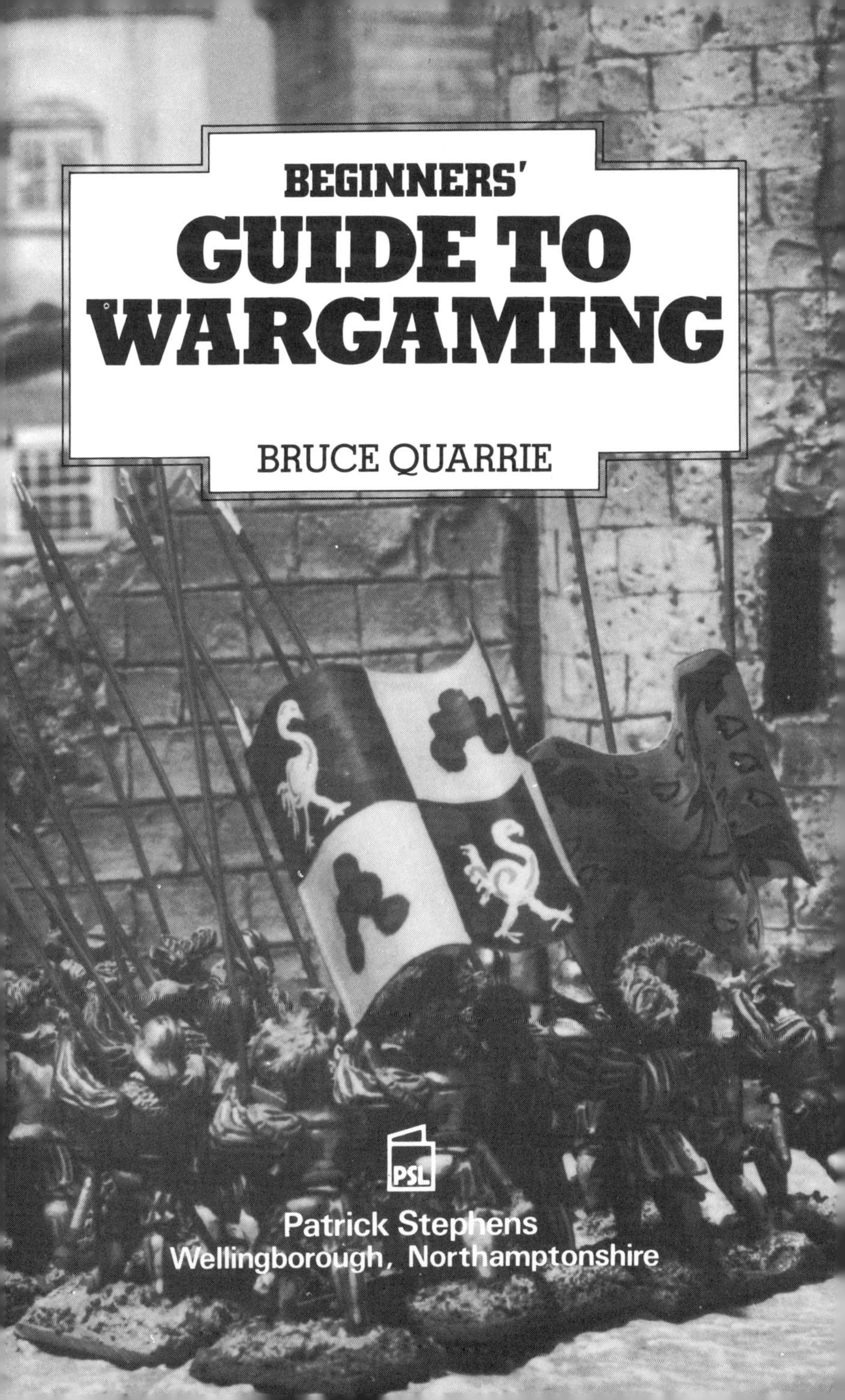

BEGINNERS'

GUIDE TO WARGAMING

BRUCE QUARRIE

PSL

Patrick Stephens
Wellingborough, Northamptonshire

First published in 1987

British Library Cataloguing in Publication Data

Quarrie, Bruce
Beginners' guide to wargaming.
1. War games
I. Title
793'.9 U310

ISBN 0-85059-852-4

Front cover picture *Agincourt period battle between English men-at-arms and longbowmen matched against the flower of French knighted chivalry* (figures by Bill Lamming, photo by Brian Monaghan). **Back cover picture** *Lansknechts, hussars and other assorted colourful troops in a 'pike and shot' period wargame* (photo by John Wylie).

*Patrick Stephens is part of the
Thorsons Publishing Group*

Printed and bound in Great Britain.

CONTENTS

Acknowledgements

Various books referred to have been invaluable in the preparation of this volume, notably *Air Battles in Miniature, Sea Battles in Miniature, Tank Battles in Miniature, Skirmish Wargaming* and the *PSL Guide to Wargaming*, all published by Patrick Stephens. For photographs I am particularly indebted to Brian Monaghan and John Wylie, and for incidental but important pieces of specific information to Paul Beaver and Terry Gander. Phil Barker also deserves a special vote of thanks. To anyone whose name I have missed, my thanks. To you, my reader, my thanks also for reading this far, and I hope you will go on to enjoy as many happy hours sweating over the fortunes of your model soldiers as I have.

INTRODUCTION

It is a sobering thought that it is a quarter of a century since wargaming in the modern sense was effectively 'launched' with the publication by Stanley Paul in 1962 of Donald Featherstone's book *War Games: Battles and Manoeuvres with Model Soldiers.* A great deal has happened in those 25 years, but there is not a wargamer alive who does not owe a great deal, directly or indirectly, to Donald Featherstone's influence.

Most boys, and many girls, play happily with toy soldiers and tanks, or cowboys and indians, or today, I suppose, *Star Wars* figures and accessories, but this is a phase which used to end with the onset of adolescence and a feeling that it was 'childish'. Today, that is significantly less true because many senior schools in Britain and America have organized wargames clubs which meet on a regular basis. Here, people can meet others with similar interests and learn the enormous variety of possibilities which modern wargaming can offer.

This book, therefore, is primarily addressed to those people without access to a wargames club or those who want to learn the basics of the hobby before 'taking the plunge', whether at school or at an adult club. There are clubs in most large towns which meet on at least one evening a week in a community centre or church hall, for example, and you can usually obtain details from your local library. Most clubs welcome newcomers and you will normally find yourself invited to take part in a game or two before anyone suggests you ought to pay your membership fee — though even this is usually remarkably low and is just designed to cover the hire of the meeting room plus the acquisition of playing tables, 'terrain' and figures for the mutual use of members.

The main point which should be apparent from the above is that wargaming is no longer the solitary pastime it used to be, nor is the hobby today regarded as any more 'eccentric' than an interest in model railways, to draw an obvious parallel since both pastimes are demanding in very similar skills.

The origins of wargaming

Where or when wargaming began is impossible to say, but it is probable that our cavemen ancestors devised some form of game played with stones similar to those still popular in some parts of Africa. Organized wargaming dates back at least 5,000 years, though, to the Chinese game of encirclement which is still popular under the name 'GO'. Throughout history, enlightened military commanders, such as Alexander the Great, are known to have used relief maps and markers to help in planning their operations. Other games such as Chess, or Christopher Weikhmann's 'King's Game' invented in 1644, really bear little resemblance to actual warfare, and the first true wargame dates back to the 18th century when Helwig, Master of Pages to the Duke of Brunswick, devised a game played on a gridded map with counters representing the three main arms of an army of the period — infantry, cavalry and artillery.

Then, in 1811 — while the Napoleonic Wars were still raging — a game invented by a Herr von Reisswitz was introduced to the Prussian court. Called *Kriegspiel*, which means 'wargame', it still used counters to represent military units but was played on a sculpted relief map and used realistic movement rates based upon the actual marching rates of armies of the time. It rapidly gained popularity and was soon being used as a method of helping to train junior officers not just in Prussia but also in Russia and Turkey.

Throughout the 19th century von Reisswitz's game was constantly discussed and modified by its adherents, who rapidly polarized into two schools of thought. One group adhered to von Reisswitz's concept of a set of rules which determined every aspect of play and from which the results of engagements were calculated. However, because no set of rules, no matter how logically or thoroughly constructed, can cover every eventuality of warfare, the rules themselves gradually became longer and longer, increasingly cumbersome and unwieldy. To break away from this another group of players tried discarding rules entirely other than for basic necessities such as march rates and weapon ranges, and used an umpire to adjudicate on the results of engagements based on his knowledge of historical precedents. In modern wargaming, a combination of both systems is usually employed, the umpire taking decisions on situations not covered by the rules, resolving disputes between players and — it has to be said — preventing cheating!

After the Austro-Prussian and Franco-Prussian Wars of 1866 and 1870 the *Kriegspiel* became increasingly sophisticated, actual statistics from the battles of these conflicts being used to modify the playing rules and bring them in line with contemporary weapons (the first machine-guns in Europe were used during the Franco-Prussian War) and tactics. As a result, both the German Schlieffen Plan of 1914 and the initial British response to it were based upon strategies which had been pre-played as wargames, and all subsequent military campaigns have used a greater or lesser degree of wargaming in their planning. Today, of course, sophisticated computers and combat simulators take the 'donkey work' out of military wargaming, enabling the participants to concentrate on decision-making rather than interpreting the rules.

Up to this point wargaming had been exclusively a military preoccupation, but in 1913 the novelist H.G. Wells wrote a super little book — which has subsequently been reprinted and is available again today — called *Little Wars*. In this he drew up rules for a game played using the popular Britain's hollow lead model soldiers and matchstick-firing guns of the period. The horrors of the Great War made wargaming an unpopular hobby, though, and Wells' book languished until its modern revival. Wells' 'little wars', however, were games pure and simple, bearing no relationship to real warfare, and the delightful cartoons with which his book is illustrated make this point quite clear. Hence we return to 1962 and Donald Featherstone's *War Games*, which produced a revolution in the hobby.

Basic principles of wargaming

Wargames fall broadly into two categories — those based on historical actuality and those based on fiction, usually science fiction or fantasy. However, the principles of rule construction are the same for both so the following general remarks are equally applicable.

A wargame is an attempt, using counters on a map, model soldiers on a sculpted terrain, or computer-generated images on a television screen, to reproduce without actual bloodshed the warfare of a given period on the land, at sea, in the air or, today, in space. The basis of any wargame is a set of playing rules — but don't be put off by the word, because many are as simple to learn as those necessary for a game of 'Monopoly'.

Any set of rules has to cope with a wide variety of factors. What troop types are involved? Are they armed with spears and bows, with swords and muskets, with machine-guns and tanks or with lasers and space ships? How fast can they move, depending on the nature of the countryside, the weather and other factors? What are their weapon ranges? What advantages or disadvantages does one type of soldier or weapon have over others? And finally, how do you simulate combat, produce 'casualties' and determine which side wins?

The first, and most essential concept to grasp is that of *scale*. A map has a scale. It will be gridded in inches or centimetres, with each unit representing so many miles or kilometres, so you can look at a map and say 'if a car travels at an average speed of 40 mph it will move such-and-such a distance over the map every five minutes, or ten minutes'. Exactly the same principle can be applied to a marching foot soldier, a cavalryman, a tank, a warship, an aircraft or a spaceship once one knows how fast they can move. Movement rates are obviously variable. The car in the above example will have a far higher average speed if it is travelling along a motorway than it will along a country lane. Similarly, a sailing ship will move faster with the wind than against it, an aircraft faster in a dive than in a climb or a space ship faster in 'hyperspace' than in normal space.

Thus, depending on the nature of the warfare you wish to recreate, the first thing to determine is a scale appropriate to the average speed of movement of your forces. A wargames table for any historical period up to the First World War, when mechanical transport was used for the first time on a large scale, will need to represent an area, in most cases, no more than a mile or two on each side, the exact scale being determined by how far infantry, cavalry and artillery can move in a given space of time. When employing tanks and aircraft, for example, the scale will have to be different because they not only move far faster but also have weapons capable of firing over greater distances.

For the sake of convenience, and because a majority of wargamers principally fight pre-20th century battles, I shall use the popular 'horse-and-musket' period as my example in explaining the basics of playing rules here.

Two yardsticks are used in determining scale. The first, as discussed, is how quickly troops can move; the second is the maximum effective range of their weapons. A typical infantryman, marching in formation across a battlefield, we could say covers a hundred yards or metres a minute, and a smoothbore musket has an effective range of about a hundred yards too. We could therefore say that an inch or a centimetre on the wargames table *represents* ten yards, so the infantry can move ten inches or centimetres in a minute and their muskets similarly have a range of ten. Horses, of course, can move faster, and we could say, for example, that cavalry can move twenty inches or centimetres in the same space of time.

Having established a basic scale along these principles, you can then modify it to allow your troops to move more quickly along good roads or when charging into contact with the enemy, and more slowly through woods or up hills, for instance. You can go even further, and give different basic marching rates for well-trained and disciplined units such as Guards regiments and for poorly trained and undisciplined units like militia or conscripts. The permutations are endless if you wish to make them so, and in one of my earlier books, *Napoleon's Campaigns in Miniature*, there are different characteristics not just for each of the various armies of the Napoleonic Wars but for each troop type within those armies.

The ground scale adopted has another function too, in determining *unit frontage* and *depth*. From historical records, we know that a regiment drawn up in line of battle or column of march will occupy a certain amount of ground, and we can therefore apportion our units the relevant space on the battlefield. For example, you could say that a battalion of 600 men in three ranks occupies a *frontage* of 200 yards (in practice slightly less) and has a depth of three yards. Unfortunately, this is where things start becoming a little complicated.

If you have seen any of the model soldiers used by wargamers, you will realize that each of them stands upon a base which can be anything up to a centimetre wide and, in the case of cavalry horses, up to five times that in depth. This depends upon the scale of the figure (from now on the word *figure* means one model soldier), which is covered below but, basically, it means that you cannot possibly cram 200 model soldiers of any reasonable size into a frontage of twenty inches or centimetres, and even if you could, by using what are called 'flat' figures, they would still occupy too much depth.

What we do, therefore, is to say that one figure, or model soldier, *represents* a certain number of actual men. Thus, if we continue with the hypothetical battalion of 600 in three ranks, each rank comprising 200 men, and have a ground scale of a centimetre to ten yards/metres, then twenty figures could represent the 200 at a 10:1 (ten equals one) ratio, since those twenty figures would occupy a scale 200-yard frontage. If this leaves you confused, draw up a strip of paper or cardboard 20 cm long and divide it into 10 mm intervals. Don't bother about the depth for the moment. A model soldier will stand upon each interval, thereby 'representing' ten men. The twenty become 200. However, the bases upon which the figures stand are also a centimetre or more in depth — representing ten yards, whereas we only want three. Thus, 'rounding' the figures, we make our twenty model soldiers represent 600 real men in three ranks at a ratio now of 30:1. This still leaves an anomaly in the depth of the formation, but it is an unimportant one — fortunately — since there would always be a slight gap beween battalions or regiments on the battlefield, and these few yards are not significant.

What *is* important is the unit's frontage, since this determines how many muskets (or bows, or pikes or spears, or whatever) can be brought to bear at any given moment. With our hypothetical battalion of 600 men in three ranks, one rank can fire from the kneeling position and the second from the standing, while the third is reloading. Thus, when we come on to calculating the effect of musket fire, the three-rank battalion will be at a disadvantage when faced by a similiar-size unit drawn up in two ranks, which can disgorge all of its weapons instead of only two-thirds... as in fact happened when British infantry in two ranks met French in three throughout the Napoleonic Wars.

With cavalry the problem of depth is much more acute, and the player can adopt one of two courses: the easiest of which is to ignore it since frontage is far more important. The alternative is to assume for the purposes of the game that any cavalry regiment is always proceeding in ranks which are widely spaced apart, and in fact there is considerable historical precedent for this since the men in the second rank would want to allow their horses sufficient room to gather themselves for a jump if the horse in front went down. This, at least, is the rationale I have always adopted although you can play with alternatives such as assigning cavalry a different man to figure ratio to that of the infantry. Unfortunately, such ideas usually lead to even greater 'book keeping' than normal during a game, slowing it up and leading to increased acrimony between players.

The same applies even more to artillery than it does to cavalry. Take a model cannon, and you will see that its width is about the same as three infrantryman, but it is much deeper — especially when you add its towing limber, horses and crew. *Frontage*, fortunately, is easy to take care of. Just as you say that one infantry or cavalry figure represents a larger number of actual men, you say that one model cannon represents a whole battery of (average) eight guns, each requiring a ten-yard front, so you mount the cannon on bases 8 cm wide (6 cm for a six-gun battery, 12 cm for twelve, and so on). This is fine when the guns are deployed for firing, but what about when they are being moved, with the combined length of gun, limber and horse team occupying ten times the ground space they should? Answer — ignore it. The model horses, etc, are present for purely decorative purposes, since a cannon would look decidedly odd moving itself!

Wargaming, you will have gathered by now, is often a matter of compromise, and it is in this respect that deciding what is important to the playing rules and what can be discarded without ruining either the game or historical accuracy becomes crucial — and the subject of much heated debate because it is, or can be, very much a matter of personal opinion depending upon which books you have or have not read.

I hope by this time that the concept of ground scales is clear. I have deliberately not discussed weapon ranges because they are irrelevant other than in their proper historical context, see later, but so long as you understand that a military unit occupies a specific portion of ground, that model soldiers can be used to represent greater numbers of actual troops, and that march rates are related to a ground scale allowing troops to move a specific distance in a given time interval, then you are half-way to understanding how a wargame works.

The next obstacle to overcome is figure scale. Depending upon the manufacturer and the scale adopted, a model soldier used for wargaming can be anything from 5 mm to 54 mm in height, although the commonest scales are 15 and 25 mm. When I started wargaming, 25 mm was by far the most popular scale, not least because the inch-and-a-bit high figures are so attractive in their own right. Due to laws limiting the amount of lead used in the manufacture of such figures, however, their price has rocketed in the last few years making the 15 mm scale models much more attractive for large-scale battles — though not for skirmishes or fantasy games, of which more later. The smaller figures have a further advantage in addition to their comparatively low price — their size enables you to have more figures in a unit, enhancing its appearance of authenticity; or more units of the same size, permitting greater battlefield flexibility.

Whichever of the many scales available you ultimately opt for, unless you decide that either skirmish or fantasy role-playing games are your forte you will quickly notice a discrepancy of scale in that, if a centimetre on the ground represents ten yards or metres, a 15 mm figure is taller, proportionately, than the average house, and a 25 mm figure the height of a block of flats. Don't worry! All it requires is a flexible imagination, and if you have come so far as to allow one model soldier to represent ten or thirty real ones, then it is an easy hurdle to surmount. Height is unimportant, since it is the visual effect of hordes of beautifully painted figures marching in perfect order across the battlefield which is one of the principal joys of wargaming.

Height, of course, is not unimportant in a tactical sense, since you will wish to position your guns on the tops of hills to add to their range and allow them to fire over intervening obstacles, including your own troops. Similarly, you will wish to put some of your own troops behind hills or hide them in woods or buildings — and since the normal hill on a wargames table is constructed from layers of polystyrene ceiling tiles cut to shape and contoured, a 25 mm figure is often taller than the 'hill' behind which he is sheltering. So, we have to choose a vertical scale for terrain features which will relate to the ground scale rather than the size of the figures. The usual way in which this is done is to assign an arbitrary height of fifty feet to all trees, houses and single contours or hills. A 'three-storey' hill of ceiling tiles is thus 150 feet high, even though in practice a model cavalryman may be able to 'look' over it.

The only ways over the 'vertical scale' problem are to use extremely tiny figures a mere 5 mm high which are usually sold cast together in battalion-size blocks, and which have no individuality whatsoever; or to use cardboard markers cut to the correct frontages and depths for each unit, in which case you might as well play upon a map as on a table. There *is* a third alternative, but it is beyond the reach of most people, and that is to make your ground scale proportional to the size of your figures ... but that means you would need to have the same number of figures in a battalion as there were in real life, and would require a large garden in which to fight a battle. (Many wargamers dream of doing this and it *is* possible, if enough people group together for a weekend during the summer and someone mows the lawn sufficiently short!)

Having established a ground scale, a vertical scale and the size of figures you wish to collect, weapon *ranges* become unimportant because they will be whatever they were historically, scaled down to your chosen level. Rates of fire, just like rates of march, are another matter. In a very simple game it might be adequate to rule that a unit can move, or fire, once every turn, but not both simultaneously or in succession during a turn. That this is unrealistic can be seen very quickly from the following example. During the Napoleonic Wars, an experienced infantryman could load, fire, reload and fire his musket again in the space of a minute. Thus, the playing rules should either allow him to fire twice in a one-minute turn; move half-distance, load and fire; or move the full distance and still fire *if* he had loaded his musket in the previous turn. (Firing on the move belongs to a later technological age and will be discussed separately.)

Now, how do you assess the casualties upon the enemy unit which result from such fire? In the early days of 'modern' wargaming a very arbitrary system was employed in which you usually threw a single dice, numbered one to six, for every twelve figures in a position to fire, and the result was the number of casualties. You can easily see that this was both unrealistic and reduced the wargame to a simple game of chance, since a player throwing a high proportion of fives and sixes would obviously beat an opponent who could only manage ones and twos. In the mid-1960s, therefore, various groups of wargamers began searching for a more realistic system and, as the military wargamers had done after the Franco-Prussian War, turned to analyses of actual battles in order to determine what level of casualties could be expected during a particular type of engagement. In the Napoleonic period, which happens to be my personal favourite, various tests were carried out by the French, Prussians and British in order to determine how accurate the musket was. These produced remarkably high results incompatible with actual casualty rates on the battlefield. Further study showed that a variety of other factors came into play, including the fact that a moving body of men is not the same as a stationary target, that the enemy would be firing back, unnerving many soldiers and causing them to fire wild, that muskets suffered a high proportion of misfires, that smoke from the black powder used caused a real 'fog of war' obscuring the target, and so on. Eventually, I was able to draw up a table giving infantry a percentage chance of hitting their target depending upon range and a variety of other factors, and this is how most firing works in modern wargames.

A dice is still used to introduce the element of luck present in every battle, but its effect is greatly reduced. (In a home computer game, the computer itself assesses the percentile chance of scoring 'hits' and throws in a random number for a similar result.) This element of chance in wargames is sometimes criticized, but without it miniature battles would become as stylized as a game of Chess and would be far less representative of real warfare.

Physical hand-to-hand combat between opposing forces using spears, pikes, swords, axes, bayonets or whatever is resolved in a very similar way.

Returning again to the early days of modern wargaming, one anomaly which concerned many players was the fact that units on the wargames table would always fight to the last man. Now there are, of course, historical precedents for this, such as the Texans at the Alamo or the French Foreign Légionnaires at Camerone, but they are extremely rare and usually only occur when a force has no line of retreat and is not permitted to surrender. So, how to make inanimate figures with no actual feelings of either courage or cowardice 'react' in a realistic manner became the next subject of scrutiny.

For a long time the so-called 'fifty per cent' rule held sway, although much debated in the magazine *Miniature Warfare* which John Tunstill ran in the 1960s, as well as in Donald Featherstone's *Wargamers' Newsletter*. (Unfortunately, neither magazine exists today, but there are replacements — see Bibliography.) The fifty per cent rule stipulated that once a unit had been reduced to half its original strength it must turn tail and retreat, taking no further part in the battle. It was a step in the right direction, but it did not take account of the fact that total casualties in any battles prior to the mass slaughter of the First World War rarely exceeded twenty per cent. Further analysis showed that there are many reasons why one unit might stand despite suffering over fifty per cent casualties, while another could run at the first shot. The physical condition of the troops, their training and discipline, the character of their officers, whether they were in a secure position or isolated and many other factors could be seen to come into play, and gradually sliding scales came into use like those for assessing the effectiveness of firepower.

Called *morale tables*, these became the principal means of resolving the outcome of battles. A carefully constructed morale table can produce very realistic results, causing assaults to waver and halt, causing spontaneous counter-attacks (often against the wishes of the player!), and even 'domino effect' situations in which one unit turns and flees and others, witnessing this, decide to do likewise even though they may not be under any immediate threat.

Simultaneous play

One of the things which most confuses newcomers to wargaming is that, under most sets of playing rules, all movement and firing is simultaneous. In other words, unlike in Chess or virtually every other game, the players do not take turns. How this works is as follows. At the beginning of each turn, the opposing players jot down on a pad of paper their orders for each unit under their command. These can be abbreviated to a form of shorthand to save time — for example, 'advance half a move, stand and fire' could become 'A½F'; 'charge at the nearest enemy unit' could just become 'C'; 'stand and await further orders' just 'S', and so forth. So long as the abbreviations are understood by both players, and the umpire if there is one, in order that there can be no accusations of cheating when a player makes an unexpected move, this system works very well. However, it does take time and you will probably find, once you have a regular opponent, that all that is necessary is to draw up a master set of orders for the conduct of the battle right at the beginning. Morale rules will take care of most of the rest, and the only time you will need to issue revised orders will be if your opponent does something unexpected or if you yourself suffer an unexpected reverse and need to alter your tactics.

One of the biggest advantages of simultaneous over alternate movement can be seen clearly from the following example. If you have two gun batteries firing at each other, each crewed by, say, five figures, one would obviously be at a disadvantage if the other was allowed to fire first. Casualties would be removed and the second gun to fire would then be operating at reduced effectiveness. With simultaneous movement and firing, *calculating* the casualties can be done alternately so that each player can follow the other's calculations and ensure no factor has been omitted, but no figures are removed until the end of the game turn.

The wargames table

Assuming that, having read this far, you have decided that you want to try playing wargames using model figures as opposed to counters on a map, the first thing you will need is a wargames table. If you have a table tennis table or something similar this is ideal, but most people don't. Nor is there room in many houses for a permanent table to be set up. What most people do, therefore, is buy two pieces of sturdy chipboard, three to four feet square and ⅝ or ¾ of an inch thick, which can be hinged together to set on trestles or on top of the dining room table (put a cloth or some newspapers down first to prevent scratches!). Painted green, this is your basic baseboard. Baize or felt could be used instead of paint, but either is more expensive, of course.

However, battlefields are rarely completely flat, so you will need some hills. These can be properly sculpted from wood, moulded in plaster or made from papier mâché, but the big disadvantage of all these methods is that, unless the slopes are *very* gentle, most wargames figures will not stand on them but will topple over. Most people therefore use polystyrene ceiling tiles cut in irregular shapes, painted green (use emulsion, not gloss paint which eats the polystyrene away) and laid on top of each other. This gives a series of flat steps on which model soldiers will stand perfectly.

Road and river lengths, rather like stretches of model railway track, can be purchased, vacuum-formed out of plastic, or you can make your own from cardboard. Many players simply use lengths of blue and brown cloth or felt, while others just use chalk to draw in the outlines of these features. However, I feel it is worth the effort to try to make your terrain features look reasonably realistic after all the time and energy you will have spent in painting your wargames figures.

Model houses, churches, farm buildings and the like can be purchased in plastic or cardboard kit form from model railway shops, and a wide variety of ruined buildings and fortifications, gun emplacements, etc, cast in resin can be bought through one of the specialist companies which advertise regularly in magazines such as *Military Modelling*. (This magazine is also an excellent guide to the many different ranges of wargames figures available from several manufacturers.)

Model trees, hedgerows, walls, fences and the like can also be obtained from model railway stockists, but tend to be disproportionately priced and unless you are wealthy it is best to make your own. Trees are easy. Collect a handful or two of suitably sized twigs, cut the end which is to form the 'trunk' flat and glue it to a piece of cardboard so that it will stand up. Then take a sheet of foam rubber and pull it into irregularly shaped lumps. Glue these to the 'branches' and spray with matt green paint. To enhance the overall effect, dried tea leaves can be sprinkled over the rubber while the paint is wet. Hedgerows can similarly be made by 'plucking' strips of foam rubber to give an irregular appearance and painting them in various shades of green and brown. Walls can be made in strips from Plasticine. Sealed with Banana Oil, Plasticine hardens and can be painted. If you can't find Banana Oil in a local shop, ordinary polystyrene cement as used by plastic modellers can be smeared over the Plasticine to produce a hard surface.

If you have both sufficient storage space in the house, and the inclination, far more realistic battlefields can be assembled using terrain modules. Here, the industrious wargamer constructs interlocking two-foot squares of terrain which can be pushed together in different configurations to give practically endless varieties of battlefield. Each terrain module is properly sculpted, with a gently undulating surface covered in flock or sawdust painted green to represent grass or grooved and painted to represent ploughed fields. Hedges, streams, roads, ponds, bridges, ditches and buildings can be permanently sculpted into place, each module in effect being a miniature landscape diorama just waiting for the model soldiers to bring it to life. (Many useful tips on terrain construction can be found in Michael Andress' book *Scenic Railway Modelling*, also published by PSL.)

Although you still need a playing surface, forms of wargame other than land battles require far less effort in terms of scenic modelling. For naval encounters all you need is a table painted blue-green with white flecks to represent waves; for aerial encounters a pale blue board on which irregular pieces of polystyrene ceiling tile can represent clouds. For space battles a simple blackboard will suffice, and stars and planets can be drawn in with chalk.

Assembling and painting an army

Wargames figures are expensive, make no mistake about it, and one of the principal advantages of joining a club is that you will be able to play with other people's figures while you are gradually building up your own collection. Figures also take a fair while to paint to any reasonable standard, and even if you can afford it I don't recommend rushing out and buying a whole army at once, because the thought of the boxes of figures still waiting to be painted can be offputting! Buy a few at a time and paint each unit as an identity, so that when you mix colours you can be sure that each figure will look alike. A gun battery, a squadron of cavalry, a battalion of infantry, these are the sort of unit sizes you should aim to tackle at a time. In more modern periods, similar quantities of tanks, aircraft or space ships should be tackled in batches.

You *can* use 1:76 or 1:72 scale model aircraft and tanks, together with their accompanying figures, for wargames. These approximate to '20 mm' figures. However, these days it is more usual to find people playing with 1:300 scale models. This scale reduces something like a Second World War German Tiger tank to about half an inch in length and gives a Spitfire a wingspan of around an inch. These tiny models can nevertheless be painted to perfection, and their small size means that their speed of movement as well as the ranges of their weapons can be reproduced in an acceptable wargames scale. Warships, whether of sea or space, again demand different scales. You can fight a very satisfactory one-to-one battle in 1:72 scale using Greek biremes and crews, but for anything more modern — or for battles involving whole fleets of ships — it is more usual to use tiny models of 1:600, 1:1,200 or even 1:2,400 scale. This makes sense when you consider that a Second World War battleship had a main armament capable of firing shells over fifteen or twenty miles, and torpedoes could be aimed with reasonable accuracy over half that distance. When one enters the realms of modern warfare the problem is accentuated even further, and in science fiction games widely variable scales are necessary if you want to use model space ships over interplanetary, let alone interstellar, distances.

The one area of wargaming in which figures tend to be larger rather than smaller is that of the 'role-playing' games, in which each player assumes the role of an individual model soldier. There are several popular varieties of such games, ranging from 'cowboys and indians' in which each player assumes the identity of a character such as Wild Bill Hickock or Crazy Horse, to the increasingly popular 'fantasy' games based upon escapist fiction in which players become warriors or wizards, thieves or clerics, and battle against each other and horrendous supernatural forces for fame and fortune. In these games, anything up to 54 mm scale figures can be used (the same size as the old Britain's ones which now fetch fabulous prices at auction), although the most common is still 25 mm.

Whatever your chosen historical or fictional period in which to wargame, painting the models can either be a pleasure or a chore. If you enjoy painting the tiniest detail of embroidery or lace on a 25 mm soldier, or millimetre-high serial numbers on a 1:300 scale aeroplane, fine! I do, and that is why my own army is so small after so many years. Most wargamers, however, adopt the attitude that, if a figure or a model looks reasonable to the casual eye, then it is all right. What is not generally acceptable, and will make you unpopular in club play with members who *do* care, is slovenly 'get them painted as quickly as possible' workmanship. If you really cannot be bothered to paint your models properly, though, because it is the gaming not the modelling that interests you, then there are plenty of people who advertise in the modelling magazines who will be happy to take your figures on — for a price, of course.

Gamesmanship

Nobody is perfect. No-one can paint figures to the standard they would really like to see, no matter how skilled they are or how much time they spend on the task. Similarly, no-one can win all the time in wargames — but some like to more than do others. So, if you are reading this book as your introduction to wargaming, think about what sort of player you might be. Is the game itself sufficient challenge, the sight of the lovingly painted models on the table sufficient reward regardless of victory or defeat — or do you want to WIN? If the latter is the truthful answer, then I suggest you play carefully if you want to keep opponents as well as friends, for no-one is more unpopular or more likely to lose both than the wargamer who mercilessly exploits every loophole in the playing rules, queries every umpire's decision, re-calculates every opponent's casualty or morale results or insists upon firing every last musket at extreme range regardless of the fact that it wastes time and has minimal effect upon the outcome of the game.

It is surely more important to enjoy a game, and share that enjoyment with others, than to win at the expense of losing friends or potential friends. Look at how petulant young children become if they don't get all their own way, and how unpopular they can become — and make sure it doesn't happen to yourself! Play hard, of course, but play fair, don't try to take advantage, point out your opponent's mistakes as often when it is to your own disadvantage as when it is to his, and you will be ensured plenty of people willing to play with you. Ignore this advice, and you will soon find yourself a solo player.

LAND WARFARE

For the sake of convenience I have broken this chapter into four periods, and at the end of the chapter you will find a set of playing rules which will enable you to get the feel of how wargames work, although the rules will vary in detail from wargames club to club. The four periods are: ancient and mediaeval; pike and shot; horse and musket; and rifle and machine-gun. Mechanized warfare is covered in the next chapter, since it is so different. However, although weapons change, the essential character of the infantryman or cavalryman, the weight of equipment carried, and the speed of movement remain essentially the same whether you are fighting a battle in 1815 BC or AD, so once the ground rules are established, so to speak, this chapter will principally concentrate upon changes in weapons technology and the effects those have had upon tactics.

Ancient and mediaeval

Basically this is the period, the longest in human history, from earliest times to the introduction of gunpowder as a significant weapon on the battlefield. The ancient Chinese knew about gunpowder, of course, and used it in rockets which no doubt scared the living daylights out of primitive tribesmen when they first saw them. Similarly, in mediaeval times gunpowder was certainly used in primitive cannon and hand guns from about 1320 but such weapons did not become really effective until towards the end of the 15th century, so for most practical purposes can be ignored.

Battles throughout the ancient and mediaeval period were principally decided by hand-to-hand combat using swords, spears, axes, pikes, halberds and similar weapons. Missile weapons such as bows and crossbows, slings and javelins were also used, of course, but because most warriors wore armour or carried shields (or both), such weapons were rarely decisive until Edward I introduced the Welsh longbow into the English army. However, the effectiveness of missile weapons cannot be discounted entirely because they could sufficiently weaken an enemy force as to ease the task of the troops taking part in the main assault.

The protection afforded by armour and shields also slowed the rate at which casualties mounted on the battlefield, and most wounds were non-fatal until one side began to break and run, thus exposing their unshielded backs. It is this which accounts for the fact that casualties among the victors, as shown by contemporary

Ancient Britons with chariots face up to a much stronger Roman force (figures by Phil and Sue Barker, photography Brian Monaghan).

battle reports, were invariably only a tiny fraction of those inflicted upon the vanquished.

The charge, by armoured cavalry, elephants or chariots, was usually the decisive factor in an ancient battle, the commander relying on the kinetic energy of a densely packed force hurtling into the enemy to knock them off their feet and disrupt their formation. Armies without cavalry, or only possessing light cavalry used as scouts and in the pursuit, would rely upon heavily armed and armoured infantry to achieve the same effect, and one of the principal delights of ancient warfare is the ability to match armies of widely differing composition against each other, permitting a great deal of creative tactical planning.

Sieges were a major component of ancient and mediaeval warfare due to the fact that most cities and many smaller towns were fortified with stout walls and garrisoned by local levies, but they tend to be unpopular as wargames because by their very nature they were often long drawn-out affairs. However, a wide variety of model siege engines, battering rams and catapults are available if you wish to try reproducing a siege in miniature.

Throughout the period under review, infantry can broadly be divided into three categories which the ancient Greeks knew as Hoplites, Peltastoi and Psiloi. Hoplites were essentially heavily

Typical medium infantry of the ancient world in quilted armour and carrying spears, javelins, swords and even a mace (figures by Phil and Sue Barker, photography Brian Monaghan).

armoured troops fighting in a tight formation with hand-to-hand weapons. They were the shock troops of the infantry and usually the force which would be principally instrumental in deciding the outcome of a battle. (Heavily armoured archers were also used in some armies, normally standing behind the Hoplites and firing over their heads into the enemy.) Hoplite-type troops may or may not have carried shields, depending on their main weapon — obviously a man wielding a double-handed axe, such as some Vikings or Saxon huscarls, would only find a shield an encumbrance.

Peltastoi were also close-combat troops in the main, but more lightly armoured and fighting in looser formation, making them more mobile. Principal weapons would be javelin, sword and/or bow or crossbow. A determined force of Peltastoi in a good position could stand up to Hoplites for a limited period of time — hopefully sufficient for their commander to bring up his own Hoplites — but they would normally be used either against other troops of their own character or against Psiloi.

Psiloi were essentially skirmishers, operating in very dispersed formation as scouts and in a pursuit. Being lightly armoured, or in fact totally lacking armour in many cases, they were highly mobile and could be used to harass an enemy's flanks, but they would have to disperse and rely on using their slings or bows from a distance when threatened by Peltastoi or Hoplites.

Next we have cavalry, who are similarly divided into three classes. First come heavily armoured shock troops, usually armed with a lance as their primary weapon and sword, mace or axe as the 'follow through' to use once the lance had been shattered or discarded after the impact following a charge. (Incidentally, it would be wrong to consider the average cavalry charge as being conducted at the gallop, as usually shown on films. Cavalry, particularly heavy cavalry relying upon a close formation to give them the maximum shock on impact, cannot retain formation or cohesion at the gallop, so a 'charge' would normally involve the horses being brought from a walk to a gradual trot, breaking into a canter a few yards from the enemy lines.)

Second are 'medium' cavalry, like their infantry equivalent capable of close formation charges or of clearing skirmishers. Less heavily armoured than the heavy cavalry but more mobile, they could be armed with virtually any weapon except the lance —

Right *The square was not new in Napoleonic times — here heavy infantry with shields and spears stand tight while their archers pour fire into the enemy cavalry and chariots during a South-East London Wargames Group ancients battle* (John Wylie).

Left *A colourfully attired member of the Norse re-enactment society which puts on bloodthirsty encounters between Vikings and Saxons at many exhibitions* (John Wylie).

spears, swords, axes, maces, javelins and bows being the most common. (Yes, mounted archers were common.)

Light cavalry were the reconnaissance and pursuit troops *par excellence*, wearing little armour and equipped with javelins or bows in addition to swords. They were fast and mobile, but would avoid any confrontation with troops in dense formation — although they would happily sweep down upon any unfortunate Psiloi who happened to be in their path and would harry the flanks of an advancing enemy army, killing stragglers and riding away with food supplies, drink and female camp followers!

Additionally, ancient armies also featured contingents of camelry, elephantry and chariotry. Camels were, as you would expect, mainly used by north African and middle eastern forces and, while slower than ordinary cavalry, had greater stamina so could move further in a day. Camels are clumsy, however, so were normally only used as transport animals or as mounts for archers. Elephants, brought back to Europe by Alexander the Great, were uncommon but were used by both Ptolemy and Hannibal, whose crossing of the Alps with them has entered legend. Their principal use was in disrupting enemy cavalry, since horses which have never experienced elephants tend to go berserk at the smell of them. However, carrying two or three soldiers armed with javelins or bows in wooden howdahs on their backs, they could also be used as 'shock troops', although a wounded and frightened elephant was usually as dangerous to its own side as to the enemy!

Chariots looked fearsome, especially if equipped with scythes on their wheels, but in practice were pretty ineffectual since they could not be charged successfully against determined troops in close formation nor were they sufficiently manoeuvrable to be of much use against skirmishers. Chariot charges were thus normally semi-suicidal attempts to break up enemy formations and create opportunities for other troops. This makes them a double-edged wargames weapon of more psychological than practical value in themselves, for while they can usually achieve little on their own except under ideal circumstances, their presence on a battlefield tends to make other players nervous and to huddle their troops together, making them particularly vulnerable to concentrated archery.

Choosing an army for the ancient period is almost as much fun as fighting with it, since the variety is enormous, and it is well worth going to a wargames exhibition or open day organized by one of the larger clubs just to get some idea of the scope. Ancient Greeks armed with long spears or later, under Philip and Alexander, with pikes; early Imperial Romans with their short stabbing spears and iron discipline; riotous ancient Britons, colourful but ill-disciplined; ferocious Vikings, forerunners of the 'Hell's Angels'; ancient Egyptians with hordes of archers and remarkably disciplined chariotry; Sassanid Persians with their super-heavy cavalry and elephants; Huns, whose high mobility enables them to run rings around everyone; the choices are legion and manufacturers are generous in the alternatives they provide. Chinese or Samurai armies can be particularly effective and are very popular too. Another beauty of this early period is the lack of uniformity in armour or dress in most armies, particularly amongst irregulars, allowing you to mix figures from several different manufacturers in one unit and paint them all in different colours.

The same applies when one crosses the somewhat arbitrary boundary between 'ancients' and 'mediaevals', which most people place in the 7th century AD, at a time when the Byzantine Empire was struggling against the Arabs who had just overrun Persia, and Europe was shaking itself down into new patterns caused by the power vacuum left after Rome's fragmentation.

Weapons during the mediaeval period were the same, essentially, as throughout the preceding centuries, although the crossbow assumed greater significance, particularly in the hands of the Genoese. The principal 'new' weapon seen on battlefields from the 1270s was the English longbow (actually of Welsh origin) which Edward I introduced. It not only outranged other types of bow in use by at least a hundred yards but, more importantly, had far greater penetrating power than anything except a crossbow. Its advantage over the latter was the speed with which it could be fired and reloaded, a trained archer being able to loose some fifteen aimed shots a minute, but its disadvantage lay in the fact that it took physical strength and prolonged training before an archer became proficient in its use. This was the main reason the longbow fell into disuse once portable and reasonably accurate firearms came into use from the late 15th century, since any idiot could be taught how to fire a musket effectively with minimal training.

The penetrating power of the longbow and crossbow was particularly important during the mediaeval period due to the advances which had been made in personal armour. Pieces of plate armour gradually began to supplement the mail hauberk which,

A clash between gaily caparisoned French knights and equally colourful Saracens during the Crusades (figures by Bill Lamming, photography Brian Monaghan).

apart from a quilted jerkin, had been the knight's sole body protection in most armies during the so-called 'Dark Ages'. Iron knee-cops and greaves replaced the old mail leggings, an all-embracing barrel helm with perforated face mask replaced the earlier skull caps or Norman-style nasal helmets, and so on. Horse armour, which had been widely used in classical times, was also reintroduced.

The mediaeval period is another rich one for wargamers, embracing such campaigns as the Norman invasion of Britain, the Crusades, the Hundred Years' War and the Wars of the Roses. Heraldry reached its peak, each knight having his personal coat of arms emblazoned on his shield, armourers constantly improved (and added to the weight!) of body armour, and new close-combat weapons resembling nothing more than can-openers came into use. Equally colourful bands of mercenaries roamed the continent, particularly during the long period of intermittent warfare from 1337 to 1453 which historians call the Hundred Years' War, and there is wide scope for wargames campaigns set anywhere from Flanders or Italy to the Holy Land. Changes were afoot, however.

An Agincourt-type battle with the English longbowmen drawn up in herce *or* harrow *formation behind sharpened wooden stakes to repulse the French cavalry charge* (figures by Bill Lamming, photography Brian Monaghan).

Pike and shot

The date of the discovery of gunpowder in Europe is not known, although the doubtful honour is usually awarded to the 13th century monk Roger Bacon. Primitive cannon are first recorded in England in 1327. These very simple devices consisted of nothing more than a hollow tube with a touch hole, firing stones or iron arrows, and for a long time there were few improvements. Edward III had three cannon with him at the battle of Crécy, but it was not until the second half of the 15th century that, mounted on their own wheeled carriages, such cannon acquired sufficient mobility to make them weapons of real significance — especially since early guns were as likely to explode in their owners' faces as to do any damage to the enemy.

Hand guns are believed to have originated in Germany, the earliest being simple iron barrels mounted on a wooden stock which the soldier held in one hand, applying a match to the touch-hole with the other. Needless to say, their accuracy left something to be desired! It was not long, however, before shaped stocks were introduced, together with primitive triggers which pulled down a lever on top of the barrel, the match being clamped to the end of this, thus leaving the firer both hands free to hold and aim the weapon. Known as matchlocks, these early muskets were very heavy and required a rest on which to lay their barrels while firing.

The introduction of effective firearms brought a final end to the feudal system, and by the end of the 17th century (which is where the 'Renaissance' or pike and shot period also ends), mounted knights, men-at-arms, mercenaries and levies had been replaced in most European countries by regular, paid armies. The introduction of firearms also led to the end of the wearing of armour other than by specialized troops such as sappers or for ceremonial occasions, since unless it was sufficiently heavy as to exhaust its wearer, armour was not proof against musket fire at the sort of ranges battles were generally fought.

At the beginning of the pike and shot period, from around 1450, musketeers could only fire very slowly, one shot a minute being fast. This made them especially vulnerable to cavalry since, apart from a sword, musketeers had no other means of defence, so the pike, averaging eighteen feet in length, came back into use. (It had, in fact, been reintroduced during the mediaeval period to protect men-at-arms from cavalry charges, but it was during the Renaissance that the pike came into its own.) Infantry units, originally known as tercios, thus became an integrated mixture of pikemen and

Above *Pikemen preceded by musketeers or arquebusiers advance across a bridge towards the town defying them* (John Wylie).

Below *Across the bridge, the Lansknechts encounter heavy mounted archers with massed musketeers behind* (John Wylie).

Above *Polish winged lancers try to stop the advance . . .* (John Wylie).

Below *. . . while the town's garrison prepares to sell its lives dearly* (John Wylie).

Above *However, the Lansknechts bring up their heavy artillery...* (John Wylie).

Below *... and their pikemen thrust through the breach created in the town wall. As for the outcome — it all depends upon the players* (John Wylie).

Above *The diversity of ancient wargaming — elephants, light and heavy cavalry, archers, slingers and spearmen in a typical colourful clash* (Brian Monaghan).

Below *The splendour of mediaeval warfare — heavily armoured knights in all their heraldic finery and archers in a typical defensive formation behind a barricade of sharpened wooden stakes* (Brian Monaghan).

Above *The Lord of the Manor and his retainers defend their home against marauding mercenaries* (Brian Monaghan).

Right *The late Charles Grant, one of the pioneering figures in wargaming, presides over a Renaissance 'pike and shot' battle at a South London Warlords exhibition* (John Wylie).

Below *The Crusades — European knights versus Saracens* (Brian Monaghan).

SOUTH LONDON
WARLORDS

Above *How it all started: in Edwardian costume, members of the Rayners Lane wargames club fight a battle using Britains' 54 mm figures to H. G. Wells' original rules* (John Wylie).

Below *Heavy artillery support pikemen and a siege tower during a Thirty Years' War assault on a walled town* (John Wylie).

A typical English Civil War encounter, in which one force is trying to capture an enemy wagon train while the other defends it. Victory or defeat in games like this is determined by the success or otherwise of the wagons in getting off the table, rather than by a massacre, and is usually enormous fun as well as requiring swift tactical judgement (George Gush).

musketeers. In the attack, the musketeers would first fire several rounds at the enemy then fall back to allow the pikemen — who still wore back, breast and leg armour, but more as protection from enemy pikes than from firearms — to advance in a dense mass with levelled weapons. If threatened by cavalry, musketeers would retire beneath the protection of the pikes, continuing to fire and reload from a kneeling position — a defensive circle presaging the infantry 'square' of Napoleonic times.

There were also changes in the cavalry during this period. Although the fully-armoured mounted soldier armed with the lance continued to be used until the end of the English Civil Wars at least, cavalry were increasingly lightly armoured and equipped with lighter lances or with pistols or, later, carbines. Similarly, there were great improvements in the artillery as gun barrels began to be cast in bronze instead of iron, becoming lighter and more mobile and permitting larger calibres.

Above *The 'real thing' — Parliamentarian pikemen advance during an English Civil War Society re-enactment. The author is in there somewhere...*

Above left *A more formal engagement involving heavy forces in a battle for a fortified stately home* (John Wylie).

Left *Troops of a slightly earlier period, probably Thirty Years' War, advance past a windmill in* tercios, *with the musketeers forming 'castles' on the corners of the pike blocks* (John Wylie).

There were many wars during this period and it is another colourful one in which to wargame. German lansknechts are a particular favourite, tough mercenaries who fought without giving or asking quarter, flamboyantly dressed in puffed and slashed costume of as many different colours as possible and armed with pikes, muskets or — a favourite weapon — great double-handed swords, often with serrated or wavy blades. Their very appearance must have been terrifying to an enemy — which was, of course, the intention! Swiss pikemen, the lansknechts' principal enemy, make another favourite army of this period, as do Poles with their famous 'winged lancers' — or Italians, French, Burgundians, Spaniards, Turks or Indians, come to that. (Detailed descriptions of all of these, with hundreds of drawings and photographs, can be found in

George Gush's book *Renaissance Armies*, also published by PSL.)

Apart from the wars on the continent, the most popular campaigns among wargamers are those set during the English Civil Wars which followed directly on the heels of the disastrous Thirty Years' War of 1618–1648 which devastated Europe. English wargamers in particular favour the Royalist and Parliamentarian armies of this period with their very able commanders such as Montrose, Cromwell, Rupert and Fairfax. It is also a period in which a wargamer can become more than usually involved because there are three large societies organized on a national basis — the Roundhead Association, the King's Army and the Sealed Knot — which stage regular re-enactments of Civil War battles up and down the country every summer, proceeds going to charity. Participating in such a battle, wearing authentic clothing (and even armour), and using weapons which are authentic except that the pikes do not have metal points, nor do the muskets and cannon fire real shot, is an enlightening experience for any wargamer. It makes one particularly critical of wargames rules which do not take sufficient account of such factors as fatigue and morale, because after an afternoon marching around a muddy battlefield, indulging in

Left *A fortified encampment during the lull before the enemy assault . . .*

Above *Musketeers of Tyldesley's Regiment of the King's Army try to storm the ramparts . . .*

Below *Heavy artillery is brought into play . . .*

A cavalry mêlée outside the stockade . . .

ferocious 'pike pushes', the most enthusiastic 'Roundhead' or 'Cavalier' can start to feel 'oh no, not again' as the enemy approaches for yet another round, and cases of regiments spontaneously withdrawing because their members have simply had enough are not uncommon.

Such re-enactments can involve several hundred — occasionally thousand — men on a battlefield, drawn together from all parts of the country, and as much care is put into building authentic fortifications and encampments as goes into the costumes, helmets and weapons. I recommend anyone to give it a try — wives and girlfriends are also welcome. (There are also re-enactment societies for other historical periods, although I have no personal experience of them. These include Romans and Vikings, mediaeval knights, Napoleonic and American Civil War forces, and even British, American and German units in Second World War dress with Jeeps and other military vehicles as well as replica rifles, pistols and sub-machine-guns. However, the English Civil War societies are the largest and put on the greatest number as well as

The infantry form 'schiltron', precursor of the Napoleonic square, to defend themselves against the cavalry.

the most authentic of scale battles.)

During the pike and shot period, the proportion of pikemen to musketeers gradually changed from approximately five to two in the early part to its reverse later on. Pre-loaded cartridges which enabled musketeers to double their rate of fire were one reason; the gradual introduction of the flintlock, which reduced the incidence of misfires, was another; but even more important was the evolution of the bayonet. To begin with, bayonets were simply swords whose hilts could be plugged into musket barrels, giving musketeers something with which to defend themselves against cavalry and also to use as a close-quarter weapon against enemy infantry. Of course, with such a bayonet fitted, the musket could not be fired, and eventually someone came up with the idea of the socket fitting, replacing the 'plug' with a ring which fitted around the outside of the gun barrel. By such a simple device was warfare again revolutionized . . .

Horse and musket

Plenty of action again here in two dramatic centuries from the late 1600s to the late 1800s, embracing the War of Spanish Succession, the War of Austrian Succession, the Seven Years' War, the American War of Independence, the French Revolutionary and Napoleonic Wars, the Crimean War and the American Civil War, although the last-named bridges the gap between this and the next section of this chapter. For the wargamer, any of the above-named conflicts provides enough material for a lifetime's reading, site-visiting and figure-painting, the Napoleonic and American Civil Wars being the most popular by far.

At the beginning of the period wars were usually limited, as in previous centuries, to the 'campaigning season' which ran from approximately April to October in any given year. Three main factors contributed to this — finding sufficient food for the men and fodder for their horses, the state of the roads, and the length of daylight, bearing in mind that armies had increased enormously in size over the previous couple of hundred years and were both slow-moving and unwieldy. Having said that, it should be remembered that armies at the beginning of the horse and musket period were small and highly professional compared with some of the conscript armies of the Napoleonic and later years.

Firepower had now become the battle-winner, the flintlock and later percussion-cap musket being able to so devastate an enemy that hand-to-hand encounters were increasingly rare except between opposing cavalry units. A well-trained infantryman in one of the Duke of Marlborough's, Frederick the Great's or the Duke of Wellington's armies could keep up a steady rate of fire of some two rounds a minute until the barrels of their muskets became clogged with powder residue, making reloading more difficult.

This period also saw the reintroduction after centuries of the skirmisher, the Greek Psiloi, operating in extremely loose formation to harass without coming to grips with the enemy, as well as the introduction of new troop types such as the dragoon — a mounted infantryman who would normally ride to battle and then dismount in order to use his carbine, but who could fight as light/medium cavalry — and horse artillery. Horse artillery was so called because, using small-calibre cannon on lightweight carriages, it could keep up with cavalry. Another new troop category which was to enjoy a brief vogue and give its name to many of the Guards regiments which survive to this day was the grenadier, armed with early and highly unreliable hand grenades whose fuzes were so inaccurate that one could be tossed backwards and forwards between opposing units until it exploded unpredictably somewhere amongst or between them.

Following the introduction of regular, paid and uniformed armies towards the end of the pike and shot period, the forces of most European countries were organized into regiments, but their sizes varied enormously from less than 1,000 to around 3,000 men. Regiments in turn were sub-divided into battalions and the battalions into companies, again of widely varying strengths and composition. Most regiments gradually came to consist principally of what was known as 'Line' infantry with a leavening of grenadier and light (skirmisher) companies or battalions.

One of the problems faced by all infantry up until the introduction of self-loading rifles was the time it took to reload. In the pike and shot period musketeers would be drawn up in formations several ranks deep, perhaps as many as twelve. Depending on whether the unit was advancing or retreating, the front rank (sometimes two ranks) would fire, and would then either file back through the unit to the rear in order to reload, or would stand while its colleagues filed past it — each rank thus firing in turn when it reached the front. Frederick the Great's infantry reached the highest sustained fire rate of all using the smoothbore (ie, non-rifled) musket at one shot every sixteen seconds, but this still left them vulnerable to charges, particularly from cavalry, for long and vital seconds. Various systems of volley fire thus came into effect, typified by that of a Marlburian battalion of twelve companies. This was sub-divided into three 'firings', so that at any given moment the battalion could disgorge one-third of its total firepower, while one third were reloading and one third awaiting their turn at the 'ready'.

Throughout the horse and musket period the principal weapon remained the smoothbore musket with a calibre, or inside barrel diameter, of around three-quarters of an inch. The most famous of all was the British 'Brown Bess' which stayed in active service for over a century, and which fired bullets weighing just over an ounce (25 grams). The firing procedure was much the same as for the earlier matchlock. The bullets, together with their black powder charges, were contained in stout paper cartridges. To load, the firer bit off one end of the cartridge and poured a little powder into the firing pan, which had a cover to protect its contents from rain and wind. The cartridge was then pushed into the barrel and rammed home tight with a wooden, brass-ended ramrod. The lock holding the flint was then pulled back to the cocked position and the weapon

The stylized warfare of the 18th century makes it ideal for wargaming, as can be seen here, but despite the efforts of Charles Grant to popularize it, it remains a minority interest (Brian Monaghan).

was ready for firing. (Flints, which were far larger than those used in a modern cigarette lighter, had to be replaced after about twenty rounds.)

Cavalry were armed with swords as their main weapon, although the lancer survived in some armies and during the Napoleonic Wars the French and Polish lancers were among the most formidable troops on the battlefield, the length of their weapons on occasion allowing them to charge infantry squares with some measure of success. Other cavalry were armed with pistols and/or carbines, but more for self-protection than for offensive use. Indeed, the Duke of Marlborough only allowed his cavalrymen three rounds of ammunition, which were to be used while guarding their horses in camp, never on the battlefield.

Right *Reload and fire at will . . . French skirmishers fall back in the face of the enemy advance* (John Wylie).

Below right *The importance of good terrain to the overall appearance of a wargame is shown in this and subsequent photos. In this encounter buildings and hills have simply been placed upon a bare and rather shiny table* (John Wylie).

Below *The English Civil War is not the only conflict to attract 'full scale' wargamers. Here, members of the Napoleonic Association in a mixture of British, French, Prussian and Austrian uniforms begin a display* (John Wylie).

Artillery was increasingly important as a decisive weapon, and a wide variety of cannon proliferated — all still smoothbore but with better and more mobile carriages and limbers. At the bottom end of the scale were three and four pounders which were used in most armies as regimental weapons, each regiment of infantry having a couple or more. Then came six pounders which were principally used by the horse artillery in batteries averaging six guns. Field

Left *The same problem again here, even though the table is matt — the chalked-in rivers and roads and the unpainted bridge do not do justice to the lovingly painted figures.*

Below *A more realistic effect has been achieved here by means of individual terrain modules, but despite being painted to represent ditches, the straight edges to the 'fields' give the game away* (John Wylie)

Above *The same problem is less apparent here due to careful positioning of trees, buildings and hedges to mask the straight lines* (John Wylie).

Above right *Strips cut from cardboard or linoleum can be laid down to represent roads and rivers, but tend to get knocked askew in the excitement of the game!* (John Wylie).

Right *In this view of a spirited Napoleonic encounter, the use of 'contoured' ceiling tiles to represent hills is immediately apparent* (John Wylie).

artillery, allocated at brigade or divisional level, consisted mainly of eight and nine pounders, while at corps or army level there were powerful batteries of twelve pounders. Napoleon was a great advocate of artillery use and frequently massed his guns in 'grand batteries' which were very effective against other armies which dispersed their guns along their whole line.

Ranges of these cannon varied from some 400 to 1,200 yards, but at extreme ranges the effectiveness of roundshot, as cannon balls

194
CAPITAL

A lot of effort has gone into this demonstration game depicting an amphibious landing by American troops against a British fort during the War of Independence in 1776. The buildings are cardboard ones sold as model railway accessories (Brian Monaghan).

were called, diminished greatly. Three other forms of ammunition were also used. First was canister — cartridges of musket balls fired in a discharge like that from a shotgun. This was devastating against dense infantry or cavalry formations at close ranges of a hundred yards or so. Then there was shell, simple round balls packed with gunpowder. These were fired from howitzers which had shorter barrels than guns and which fired on a much steeper trajectory, dropping their shells into the middle of enemy formations. Finally there was shrapnel, which was really only used by the British until after the Napoleonic Wars. This incorporated the attributes of shells and canister, dozens of musket balls being packed inside an explosive shell to hail down on the enemy with horrible results.

With the final abolition of the pike at the end of the 17th century, battlefield tactics also changed and continued to do so throughout the horse and musket period. During the 18th century most armies were still comparatively small and, particularly those of Britain and Frederick the Great's Prussia, were composed of highly trained, skilled and professional soldiers. They still had to fight in close formation in order to make their musketry effective and, because of the need to maintain perfectly straight firing lines, could only manoeuvre very slowly. The American War of Independence changed all that, for the Americans fought in much looser formations, using ground and concealment in ways foreign to their European adversaries. They also introduced the rifle on to the battlefield. Although still flintlock muzzle-loaders, early rifles, with grooves inside the barrel to impart spin to the bullets, were far more accurate than muskets and had greater range.

By the time of the Napoleonic Wars, the British army had introduced regiments of riflemen, dressed in dark green uniforms instead of the bright red and white of the other infantry, and trained to operate in loose formation as skirmishers. The French had large numbers of skirmisher regiments which proved highly effective against the armies of Austria, Prussia and Russia, and gradually all other European armies followed suit.

The Napoleonic Wars from 1796 to 1815 exercise a fascination all of their own amongst a very high proportion of wargamers and, apart from ancients, it is with an army from this period that you will be most likely to find regular opponents. It is highly colourful, with a wide variety of different troop types, and the campaigns of the period ranged from Spain to Russia. Because, for the first time, they were wars of nation against nation rather than conflicts between small armies of professionals who did little to disrupt the lives of the civilian population, the Napoleonic Wars also saw the end of the old 'campaigning season', and battles were fought in the middle of blizzards as well as in the heat of summer. These factors allow the inventive wargamer to use a wide variety of scenic effects and different types of terrain on the wargames table. Model soldiers can also be purchased in winter or summer campaign dress to add to the effect. The charisma of many of the military commanders of the period also add to its appeal, ranging from Napoleon and his brilliant coterie of Marshals such as Davout, Masséna and Ney to the Duke of Wellington and Sir John Moore, the Prussians Blücher and

Lots of troops on the table, hallmark of any Napoleonic encounter — but please get the Coke mugs out of the way, chaps! The player in the centre is measuring the range before firing his muskets (John Wylie).

Scharnhorst, the Russians Bagration and Kutusov or the Austrians Charles and Radetsky, among many others.

The size of armies also increased enormously during the Napoleonic Wars, levies of conscripts in their hundreds of thousands being called to the colours. Because their training could not be as prolonged or intensive as in the 18th century, units had to be larger and the French in particular relied on size rather than quality to win their battles. Tactics, however, remained much the same, albeit modified to take into account the greater numbers and the enhanced effectiveness of artillery. Following initial contact (unless the enemy was in a prepared position, in which case a commander had more time to plan), part of the army would be used to 'pin' the opposition, to try to hold him to his positions whilst friendly troops were deployed for a flank attack — still the most effective way of winning a battle and minimizing casualties. As this manoeuvre became revealed, the rival commander would be

An American Civil War battle with the infantry, in skirmish order, in danger of being ridden down by the enemy cavalry. The introduction of 'battle-damaged' buildings and such items as haystacks helps create a good illusion even if the basic terrain is simple (John Wylie).

forced to use up his reserves or to swing part of his line to meet the threat. Concentrated artillery fire would then be poured into those units forming the 'hinge', while the main body of the army, formed up in assault columns, would move inexorably into close contact. Frequently their opponents, weakened and demoralized by the heavy fire they had sustained, would break and run (or 'rout', to use the correct military terminology) before physical contact was made. This would be the moment at which the canny commander would unleash his cavalry, and his opposite number would have no choice other than flight or surrender.

The 'textbook battle' as described is much the same throughout history, whether the forces are armed with spears and bows, muskets and cannon or tanks and helicopters, and what differentiates a successful military commander — or wargamer — is largely his or her judgement of timing. Release the cavalry too

A larger-scale American Civil War battle for control of a river crossing. The vacuum-moulded plastic river lengths are realistically curved and have 'banks' to enhance their realism (Terence Wise).

soon and they will be decimated by the enemy's firepower. Fail to pin the opposing force successfully and its commander will be able to manoeuvre to meet your outflanking movement.

After the trauma of the Napoleonic Wars, Europe saw a period of comparative respite, although there were armed revolutions in Spain, Portugal, Italy, Prussia and Belgium, not to mention an almost continuous campaign between Russia and Turkey, while in America there were constant battles against various indian tribes and the Mexicans. However, it was the conflict between Russia and Turkey which led to the next major European war when Britain and France allied themselves to the latter's cause. It was a conflict in which horrendous casualties were inflicted, not least because most infantry were now armed with rifles fired by means of a percussion cap impervious to the weather instead of the flintlock, although in other respects — including massed cavalry charges — the Crimean War of 1854-56 was an 'action replay' of the Napoleonic period.

The American Civil War was another matter, and the years 1861 to 1865 proved the watershed between what I have for convenience called the 'horse and musket' period and that of the 'rifle and machine-gun'. The American Civil War saw the first appearance on the battlefield in really large quantities of rifled small-arms and artillery, including repeating rifles; of the earliest genuine machine-guns; of the first military use of the steam locomotive in battle and as a means of transport; of the widespread use of the telegraph as a means of quick communication; and of the first real trench warfare.

At the beginning of the war, however, the troops of both sides marched off to battle still mainly equipped with smoothbore muskets and cannon and fought in similar formations to those of their Napoleonic forbears, organized in regiments of 500 to 1,000 men. Cavalry were particularly important during the American Civil War, and both sides formed independent brigades which scoured the countryside in isolation from the main armies. The Confederacy enjoyed a marked superiority in the quality of its horses and their riders during the early part of the war, but later the balance swung in the opposite direction.

Breech-loading repeating rifles and cavalry carbines began to be introduced in 1863, the best-known being the Sharps, Spencer and Henry, although the principal infantry weapon on both sides remained the 1861 model Springfield muzzle-loader. Like earlier smoothbore muskets, muzzle-loading rifles only had a rate of fire of two to three rounds a minute, but the breech-loaders such as the Spencer and Henry could average sixteen to twenty. The cavalry used the Enfield muzzle-loading rifled carbine in the early years but this was replaced by the Spencer breech-loading carbine which could keep up a rate of fire of ten rounds a minute and is widely regarded as the finest cavalry weapon of the war. Sabres and pistols were also used, of course.

A mixture of smoothbore and rifled guns was used by both sides throughout the war, the most common being the twelve-pounder Napoleon smoothbore and the twenty-pounder Parrott rifled gun which had nearly twice the range. However, the smoothbores were more rugged and reliable than the rifled guns and were as devastating against infantry at close ranges when loaded with canister as they had been half a century earlier.

The overall effect of the new weapons was to discourage the massed infantry formations in which tens of thousands died during

Not a wargames table laid out in a grid, but a competition table at a wargames exhibition with figures and dioramas in different classes. Even if you find you always lose on the battlefield, you may discover a forté in converting, animating and painting figures for such events (John Wylie).

the early campaigns, and to encourage looser, more dispersed formations and the use of natural or man-made cover. The breech-loading rifles in particular helped this by permitting men to reload while lying down — a difficult operation with muzzle-loaders. Infantry now fought in extended two-deep lines with plenty of skirmishers in front to harass the enemy and draw fire. These lines would advance in waves two or three hundred yards apart to reduce casualties from rifle and artillery fire, but by the end of the war the science of trench warfare had become so sophisticated that frontal assaults were suicidal, foreshadowing the shape of things to come in the First World War. Machine-guns were also used in small numbers but with the exception of the Williams, which was really a small repeating cannon firing a one pound shot, were extremely unreliable until the advent of the Gatling in 1864 and even this was not very accurate in its original version.

Rifle and machine-gun

In the five short years between the end of the American Civil War in 1865 and the beginning of the month-long Franco-Prussian War in 1870, however, vast strides were made in weapons technology. The old round bullets and cannon balls had been replaced by streamlined bullets and shells with greater ranges and penetrating powers, while the Prussian Dreyse and French Chassepot bolt-action rifles greatly enhanced the infantry's rate of fire and accuracy. A mixture of smoothbore and rifled artillery was used by both sides, the Prussian guns (the first to be cast in steel) being particularly effective: a six-pounder for example having a range of 4,000 yards, four times that of a smoothbore of the same calibre. The French also introduced the Mitrailleuse, a multi-barrelled machine-gun which could fire 150 rounds a minute to a range of 2,000 yards. However, they mishandled it tactically, treating it like an artillery piece, so it did not have the battlefield effectiveness it might otherwise have done.

Strangely, while the American Civil War is extremely popular amongst wargamers, the cavalry and fluid battles of the early years permitting a great deal of imaginative manoeuvring, the Franco-Prussian War is not, and the ranges of model soldiers available are far less extensive or varied.

Finally, before entering the mechanized age, two other wargames campaigning areas should be mentioned: the colonial wars in Africa, India and China, and the First World War. The Zulu wars are popular for a variety of reasons. The contrasting figures of beplumed Zulus with shields and assegais versus the red-coated, pith-helmeted British infantry make extremely colourful wargames but, and herein lies the main fascination, this period enables one to all intents and purposes to match 'ancients' against 'moderns'. The various other battles such as those of the British against the Mahdiists at Khartoum and Omdurman, or the British in India against colourful hill tribesmen, also make good wargames. So does the Boer War, in which to begin with red-coated Englishmen pitched themselves against the highly mobile Boer Commandos before they learned their lesson, adopted khaki and learned to emulate the Boer tactics with flying columns of lancers. However, for some reason the Boer War is not popular either, although I am sure many wargamers would relish a change if you brought out two armies at a club meeting.

We come finally to the First World War, the so-called 'Great War' to end all wars. Apart from naval actions and aerial dogfights, this is

Above *Demonstration of comparative figure scales, all of machine-gun teams of the Colonial period. From left to right they are 15 mm, 20/25 mm, 30/35 mm and 54 mm* (figures by Ted Herbert, photography Rick Reeves).

Below *An elephant-drawn artillery piece of the British Army in India during the late 19th century, one of the many attractive models available for this period* (figures by Ted Herbert, photography Rick Reeves).

probably the least popular of all amongst wargamers largely, I suspect, because the usual vision is of the horrific years of trench warfare. However, it was mobile in the beginning and again at the end, after the invention of the tank. Those wargamers who do fight in this period usually choose small engagements between platoons or companies of infantry, perhaps with some cavalry or a couple of armoured cars and mortars, rather than pitched battles between whole regiments, brigades or divisions. Such engagements are often fought on a one-to-one basis too, one model soldier representing a single infantryman, rather than using a figure to represent a greater number of men. This type of wargame is usually called a 'skirmish' and requires different rules from the larger battles of earlier periods, because each figure can be assigned a 'character' and various factors such as individual training, experience, weapons expertise and courage can all be brought into play. Such skirmish games can also, of course, be fought in other periods, and really come into their own when one enters the world of fantasy, discussed later.

* * *

Having covered the basic elements of land warfare through the ages, we can now draw up a relatively simple set of playing rules which can be adapted for figures from all periods, so that you can get a 'feel' for how wargaming works before moving on to more complicated systems. However, don't be misled into thinking that a 'simple' set of rules is necessarily inaccurate in its representation of actual warfare — the reverse, in fact, can often be the case. The main advantage of a simple set of playing rules is that they permit fast and furious action, with a minimum of time-wasting calculation.

Playing rules for land warfare

1 All movement is simultaneous, as explained in the introduction.
2 Each model infantry or cavalry figure represents twenty actual men. Each model artillery piece, from ballista or catapult to rifled cannon, represents two actual pieces and is crewed by two figures. When one figure is eliminated, the weapon's effectiveness is halved.
3 Each game turn, or move, represents one minute of actual combat time.
4 Each inch (25 mm) on the wargames table represents ten paces (roughly 25 feet) and all movement and firing is measured in paces.

5 Apart from model figures and a table, you will need a tape measure with which to measure movement and firing (a flexible steel rule is even better) and two sets of dice, one ordinary numbered from one to six and the other pair being what is known as 'average' dice. These are numbered 2, 3, 3, 4, 4, 5 and can be obtained from all wargames shops, but if you can't obtain any you can overpaint the spots on a pair of ordinary dice. The purpose of 'average' dice is to reduce the disproportionate effect of similar units throwing ones and sixes respectively, which would make the wargame purely a matter of luck instead of skill.

Sequence of play

1 Write orders. As explained in the introduction, general orders for the conduct of the battle can be written out by each player at the outset and new ones only issued when necessary, or fresh orders can be written at the beginning of each turn. Orders must be written at the beginning of any turn in which you want a unit or units to do something not covered by the overall plan, such as charge or retreat.

2 Declare any charges to be made. Cavalry which have been given the precautionary order 'countercharge if charged' may do so, other units must stand and receive the charge while standing still, the exception being skirmishers who may try to run away or evade the charge at will. Once all charges have been declared, the players test the morale (see below) of all units charging or being charged to see whether they comply with their orders.

3 Preliminary missile fire. Because some weapons — bows, rifles — have considerably greater range than others — javelins, grenades — firing takes place at two stages during each turn.

4 Movement. Move any or all of your units up to their maximum movement allowance. No units *have* to be moved except those reacting involuntarily to a morale test. Units firing distance weapons in phase 3 above must stand, apart from cavalry archers or, later, cavalry armed with pistols or carbines.

5 Second firing phase. Throw any desired close-quarter weapons.

6 Mêlée. Fight any hand-to-hand encounters.

7 Morale. Test the morale of any unit(s) according to the table which follows later.

8 Return to phase 1.

Note that, since all movement and firing is simultaneous, no casualties from missile fire are removed from the table until the end of phase 5. Further casualties from hand-to-hand combat are then removed at the end of phase 6 before testing morale.

Movement

As stated above, an inch represents ten paces, so a troop type with a movement factor of 60 may be moved up to six inches. Troop types are depicted in the rules by abbreviations, as follows: SHC — super-heavy cavalry, fully armoured upon armoured horses; EHC — extra-heavy cavalry, with mail armour and partially armoured horses; HC — heavy cavalry, with partial mail armour such as a corslet; MC — medium cavalry, wearing leather or quilted armour, or unarmoured but carrying a shield; LC — light cavalry with no armour or shield; SHI — super-heavy infantry, fully armoured in plate; EHI — extra-heavy infantry, fully armoured in mail; HI — heavy infantry, partially armoured in mail or with a plate corslet; MI — medium infantry, wearing leather or quilted armour; LI — light infantry, unarmoured; HCh — heavy chariotry, those drawn by over two horses and having more than two crew figures; LCh — light chariotry, drawn by a pair of horses and just having an unarmed driver and one armed warrior; El — elephantry; Cam — camelry; HArt — heavy artillery, which will vary from period to period, so is sub-divided below for fire effect although movement rates remain the same throughout for the sake of convenience; MArt — medium artillery; and LArt — light artillery. A final addition in the later part of the period is MG — machine-gun — which for movement purposes is classed as medium artillery.

Movement rates are as follows:

Troop type	**Normal**	**Charge**	**Troop type**	**Normal**	**Charge**
SHC	80	120	LCh	120	160
EHC, HC & MC	120	160	El	80	120
LC	160	200	Cam	120	120
SHI	40	60	HArt	40	—
EHI, HI & MI	60	80	MArt	60	—
LI	80	100	LArt	120	160*
HCh	80	120	MG	60	—

*Horse artillery of the 17th century onwards only.

Charges may only be made if they have been declared (see above). Their advantage is that they confer greater impact in the ensuing mêlée.

All movement is halved when uphill or through difficult terrain — woods, marshes, towns and villages, soft sand, etc; add 25% to the movement rate if along a good, paved road (eg, a Roman road). Roads should be classified paved or unpaved before the game begins. Chariots and artillery can only cross walls through a breach or a gate. Infantry and cavalry can cross walls without a movement penalty but become unformed (see Disorder, below). Infantry in a square or similar formation to repel cavalry can only move at a quarter of the above rates.

Formation changes

Changing from one formation to another takes time which must be deducted from the maximum movement permissible per turn. In other words, if an infantry unit forms square at the beginning of its turn (half a turn), it only has half its movement allowance left — and can then only use a quarter of that in the same turn.

Move from close to extended order or vice versa — ¼ turn.

Increase or decrease the frontage of the unit by two figures — ¼ turn.

Altering front of unit by more than 30° and less than 150° (measure with a protractor) — ½ turn. (0-30° and 150-180° changes are virtually instantaneous.)

All other changes of formation — ½ turn.

There are no penalties for changing a unit's stance, eg, from 'rest' to 'fire'.

Disorder

Units may become disordered or 'unformed' under certain circumstances, reducing their fighting ability. Those circumstances are: when moving through woods or marshes, whilst inside buildings, whilst moving on steep hills or when crossing walls or thick hedges (LI excepted); if engaged in hand-to-hand combat from flank or rear, or while changing formation, and at the end of the first turn of hand-to-hand combat in any case; while routing or pursuing; or if interpenetrated while in close order (ie, a friendly unit cannot pass through the ranks of another without their both becoming disordered unless they are in extended formation; skirmishers always operate in extended formation). Finally, formed cavalry charging down steep hills also become disordered.

Unit frontages

It is important that all model soldiers be based on cardboard bases of standard widths so as to obtain the appropriate advantages and disadvantages in terms of movement, formation changes, vulnerability to missile fire, ability to interpenetrate and effectiveness in hand-to-hand combat. Another factor which makes this necessary is that the sizes of figures vary from manufacturer to manufacturer. The following base sizes per figure are recommended for 25 mm scale figures and should be halved for 15 mm.

SHC, SHI, EHI, HI and MI — 15 mm; EHC, HC, MC and Cam — 20 mm; LC and LI — 30 mm; LCh, EI and LArt — 40 mm; HCh and MArt — 60 mm; and HArt, including such devices as siege engines — 80 mm.

Missile fire

As stated above, this takes place in two phases during each turn once units are in range — first the long-range fire and then the close-range after movement has been completed. Weapon ranges are as follows; all measurements are again in paces.

		Short range	Medium range	Long range
Hand-hurled weapons and pistols		40	—	—
Slings, mounted archers, smoothbore carbines		40	80	120
Staff slings, bows, smoothbore muskets, rifled carbines		80	160	240
Longbows, crossbows, rifled muskets		100	200	300
Rifles		300	600	1,200
Machine-guns		450	900	1,800
Artillery — ancient and mediaeval		N/A	N/A	480
Smoothbore cannon:	LArt	150	300	450
	MArt	250	500	750
	HArt	400	800	1,200
Rifled guns:	LArt	500	1,000	2,000
	MArt	1,000	2,000	4,000
	HArt	2,000	4,000	6,000

The reason different range brackets are given is because the penetrating power, and therefore effectiveness, of all weapons is greater the closer the target. It will be noted that rifles, machine-guns and, in particular, rifled guns, have maximum ranges greater than the length of the average wargames table (normally eight to twelve feet) and when fighting battles from 1863 onwards a different ground scale is needed to take this into account. A common scale is one centimetre instead of an inch to every ten paces. Fortunately, it is rare to see heavy rifled guns on the table!

Another factor which must be taken into account is who can fire and at what targets, assuming they are in range. Firstly, since until modern times virtually all fire was by line of sight only, you cannot as a general rule fire over obstacles such as woods, houses, hills or intervening troops unless your own troops are on a higher level than the obstacle. Since, as explained in the introduction, houses, woods and single contours of hills are counted rather arbitrarily as being fifty feet high, you can only fire over a house or wood if your unit is on at least the second contour of a hill, and so on. Similarly, a unit on the first contour of a hill can fire over the heads of another unit on level ground, but not vice versa, although mounted archers, whether on horses, camels, elephants or in chariots, can fire over the heads of infantry. Archery fire was, however, often 'plunging', ie, fired on a high trajectory so that the arrows would drop down into an enemy formation from above, so archers (but not crossbowmen or musketeers) may fire over two ranks of intervening troops. This means that a maximum of three ranks of archers in close formation may fire; when in extended order up to five ranks may fire. In later periods, only the first rank of musketeers or riflemen may fire from close order, up to three ranks when in extended order.

All units may only fire within 30° of their front.

Each type of weapon is assigned a different numerical factor depending upon the type of target against which it is firing, since it is obvious — to take a couple of examples — that a super-heavy armoured knight would be totally impervious to a hand-hurled javelin, or that infantry in close order would be more vulnerable to artillery fire than skirmishers. These factors are tabulated on page 73, the abbreviations along the top representing the type of *target*.

Add +1 to each factor when firing within the 'close range' bracket and deduct -1 over medium range.

Further additions and deductions should be made as follows: if the target unit is carrying shields, deduct -1 unless firing into its unshielded flank; deduct -1 if target unit is lying down or in 'soft' cover (eg, in woods or behind wooden barricades); deduct -2 if target is in 'hard' cover (eg, trenches, dugouts, stone or brick buildings or behind stone walls); deduct -1 if heavy or medium infantry or cavalry are in extended order; add +1 to smoothbore artillery if firing canister (short range only). You must write this in your orders at the beginning of the turn otherwise the gunners will be assumed to be still firing roundshot and the basic factor remains the same.

Members of the Napoleonic Association in Saxon (left) and Prussian uniforms demonstrate their skills at one of the many wargaming conventions held every year (John Wylie).

Above *A Western gunfight fought using 'skirmish' rules* (John Wylie).

Left *French infantry form square to defend themselves against Russian cavalry in a typical Napoleonic engagement* (John Wylie).

Below *Members of the Pinner Wargames Club staging an accurate recreation of the battle of Arnhem in 1944 using 1:76 scale figures and equipment* (John Wylie).

Above *Dressed in Second World War GI uniforms, wargamers prepare for an engagement using radio-controlled model tanks* (John Wylie).

Below *The shape of wars to come — some of the many figures available for science-fiction wargames* (John Wylie).

Weapon (see key below)	SHC SHI	EHC EHI	HC	MC Cam	LC	HI	MI	LI Art	HCh	LCh	EI
A	N/A	0	2	3	4	1	2	3	2	1	1
B	2	3	3	3	1	2	2	1	2	1	1
C	N/A	N/A	4	4	2	3	3	2	N/A	N/A	2
D	N/A	N/A	6	6	5	6	6	4	N/A	N/A	4
E	3	3	3	3	1	3	3	1	2	1	2
F	2	2	2	2	1	3	3	1	N/A	N/A	2
G	3	3	3	3	2	4	4	2	N/A	N/A	3
H	4	4	4	4	3	5	5	3	N/A	N/A	4

A — hand-hurled weapons, slings, bows, pistols and smoothbore carbines; *B* — longbows, crossbows, staff slings, smoothbore muskets and rifled carbines; *C* — rifled muskets and rifles; *D* — machine-guns; *E* — ancient and mediaeval artillery; *F* — light artillery, smoothbore or rifled; *G* — medium artillery; *H* — heavy artillery.

We now come to the dice throw to represent the element of chance present in any battle. Irregular troops, in other words those not trained and disciplined to fight as cohesive units, use ordinary dice numbered one to six. Regular troops use 'average' dice numbered two to five. Examples of irregular and regular troops from different periods would be ancient Britons versus Romans, Napoleonic conscripts versus British infantry, American Revolutionary marksmen against Europeans, Boers against British, and so on. It is usually easy to make the distinction. Throw two of the appropriate dice for each unit firing and add *the difference between them* to the factor of the firing unit taken from the preceding table (thus a throw of two on one dice and five on the other would give +3). You will see from this that irregulars *can* achieve better scores than regulars, but firepower from regulars will be more consistent. If both dice thrown show the same score, there is no addition.

From the beginning of the 17th century, all artillery crewmen are counted as regulars.

Having arrived at your total, refer to the following table to see what the enemy casualties are. The figures represent *men*, not model soldiers, and you only remove a figure from the table when you have scored over twenty casualties on the unit. To keep track of casualties, you should prepare a list of all your units at the beginning of the battle and write down their losses at the end of each turn. If more than ten figures are firing, simply add together the appropriate numbers (eg, sixteen men firing at +5 would be either 10 at 25 plus 6 at 15 = 40 or twice 8 at 20 with the same result). Note that -1 is the lowest score permitted, although -3 is theoretically possible. Similarly, +10 is the maximum allowed.

Score	**Number of figures firing**									
	1	**2**	**3**	**4**	**5**	**6**	**7**	**8**	**9**	**10**
-1	0	1	2	2	3	3	4	4	5	5
0	1	1	2	3	3	4	4	5	6	6
+1	1	2	2	3	4	5	6	6	7	8
+2	1	3	4	5	6	8	9	10	11	13
+3	2	3	5	6	8	9	11	13	14	16
+4	2	4	6	8	10	12	14	16	18	20
+5	3	5	8	10	13	15	18	20	23	25
+6	3	6	9	12	15	18	21	24	27	30
+7	4	8	12	16	20	24	28	32	36	40
+8	5	10	15	20	25	30	35	40	45	50
+9	6	12	18	24	30	36	42	48	54	60
+10	7	14	21	28	35	42	49	56	63	70

Hand-to-hand combat

	SHC SHI	**EHC EHI**	**HC**	**MC Cam**	**LC**	**HI**	**MI**	**LI Art**	**HCh**	**LCh**	**El**
Cavalry or camelry											
Long spear	2	2	2	4	3	3	4	5	1	2	1
Sword	1	1	1	3	2	2	3	5	0	0	0
Other	0	1	1	2	2	2	3	5	0	0	0
Infantry											
Pike	3	4	5	5	2	2	3	2	3	4	1
Shorter spear or gun with bayonet	2	2	3	4	3	2	4	3	2	3	0
Two-handed sword or axe	4	5	5	5	4	5	5	3	1	1	1
Other	1	1	0	1	1	1	2	2	0	0	1

Add a further +1 to any of the above factors if fighting from a chariot or elephant.

The above table is also used in calculating casualties from hand-to-hand fighting, or mêléeing as you will hear it called by most wargamers. As with missile fire, you first look at the type of weapon or weapons being carried by your own unit, then at the type of target. (It is recommended that if you have a mixed unit, usually of irregulars, armed with several different types of weapon, you take one as an 'average' rather than doing separate calculations for each, which would be very time-consuming!) You then throw dice as for firing, and make any appropriate additions or deductions as follows: add +1 if attacking from flank or rear, if enemy is disordered and your unit is not, if you are charging and he is not, if he is falling back and you are pursuing, or if your unit is uphill of his; deduct -1 if your unit is unformed, or if yours is falling back, or if the enemy has shields and you do not.

In hand-to-hand combat, troops such as artillerymen or machine-gunners count as having 'other' infantry weapon.

Morale

As I said in the introduction, any set of wargames rules succeeds or fails according to the standard of its morale section. Battles are usually won as a result of soldiers giving up and fleeing, retreating or surrendering rather than being killed. The object of the morale rules is to let the commanding general — you, the wargamer — know when your model soldiers have had enough!

You test the morale — or fighting spirit, if you like — of your troops under the following circumstances (or more frequently if the umpire, supposing you have one, says you should; a good umpire can often see situations not covered by the letter of the playing rules which should nevertheless be investigated).

When commencing or receiving a charge; if attacked in flank or rear; if irregulars, when coming under missile fire for the first time in the game; after each turn of a mêlée or round of hand-to-hand combat regardless of losses; after losing ten per cent of the unit's strength in a single turn, or after losing 25 per cent in any case and every move thereafter; if a friendly unit on either flank retreats or routs; or if the officer commanding the unit is killed (see below).

Regular troops throw *one* average dice, irregulars a one to six dice. To the score, add or deduct according to the table overleaf, most of which should be self-explanatory. Where the sign ± occurs, add or deduct to give the most desirable result. 'General' means the unit commander — *double* these figures if applied to the overall commanding officer, Alexander, Caesar, Napoleon or whoever, for *all* units in the army. 'Fired upon' or 'mêléed by friendly unit' — well, we all make mistakes, or they might have changed sides...

Own troops advancing	+1	Own troops in cover	+1
Own troops standing	0	For every ten per cent casualties from original strength of unit	-2
Own troops retreating	-1	For each secure or insecure flank	±1
Own troops routing	-3	Each enemy unit to flank or rear	-1
For each friendly formed unit within 150 paces	+1	Own unit in 'uncontrolled charge'	+3
For each friendly routing unit within 150 paces	-1	General with unit	±2
For each enemy formed unit within 150 paces	-1	General killed	-2
For each enemy routing unit within 150 paces	+1	Under fire for first time (irregulars)	-1
Own troops disordered	-1	Fired upon by friendly unit	-2
Own troops uphill from enemy	+1	Mêléed by friendly unit	-3
Own troops downhill from enemy	-1		

The above can give a result from about -10 to +15, depending upon circumstances. If the total is a minus score, the unit throws down its weapons and runs away from the enemy at charge speed towards the nearest edge of the wargames table; if the total is between 0 and 3, the unit will waver and refuse to advance any further until its morale level is raised, usually due to the removal of a threat or the added presence of the commanding officer; if it is between 4 and 8 the unit may carry out its orders but will not charge without a further morale test; if it is between 9 and 12 the unit will do exactly what is ordered or desired. Over 12, however, a unit is assumed to have got real blood-lust and gone berserk, and it will charge straight at the nearest enemy unit, regardless of anything else! It continues in this fashion until after the first turn of mêlée, when morale must be tested again. This is the 'uncontrolled charge' referred to in the table above.

All morale points are accumulative, so that a routing unit (-3) which is also of necessity disordered (see above), starts off with -4; a couple of friendly units standing firm plus the presence of the commanding officer — depending upon the dice throw — may thus be necessary to halt its flight. If it continues running, though, all units within 150 paces must test their own morale and may follow suit.

Generals and commanding officers

Since each model soldier represents twenty real men, it is

impossible to say when a general is killed other than by resorting to a chance system. So, each time you remove a figure from the table, throw an ordinary dice. If the score is one, the unit commander has been killed. Throw the dice again. If the result is another one, the overall commander has been killed instead, *but only* if the figure representing him was placed within 150 paces of the unit concerned. To prevent players placing their officer figures 'out of reach', each must at all times be within 150 paces of at least one friendly unit. Cautious wargamers place their commander figures to the rear out of harm's way, with a bodyguard well within 150 paces — but this sacrifices the enhanced effect their presence has in the front line.

Additional rules

The preceding are, as explained, relatively simple playing rules which will produce enjoyable games. They are based upon the much more complicated sets produced for each historical period by the Wargames Research Group, who also publish army lists enabling you to select from the many available not only which army to purchase but in what proportions to buy your models so as to end up with an army of model soldiers whose characteristics accurately reflect the genuine article. WRG also publish several books on the armies of different periods which go into much fuller details of organization, weaponry and dress than this book possibly can. Further ideas which could be incorporated with ease into the foregoing playing rules though include such facts as the weather — pouring rain will play havoc with bowstrings or gunpowder; smoke — from cannon or burning buildings, obscuring your gunners' view of their targets; daylight — limiting the length of the game; and so on. You could produce separate missile and hand-to-hand weapon charts for a much wider variety of weapons, giving each subtle advantages and disadvantages against others. You could introduce variations on the 'regular/irregular' troop type theme, giving bonus points for Guards units both in their fighting effectiveness and their morale. The list of possibilities is, literally, endless, and only limited by your own reading in what finally emerges as your chosen favourite historical period.

Even if none of the foregoing appeals, though, there are many other alternatives...

ARMOURED WARFARE

The name 'tank' was originally a cover name to hide the new weapon being developed by the Landships Commission set up by Winston Churchill in 1915, but since 1917 tanks have played a crucial role in virtually every major battle around the world. The first such vehicle was christened 'Little Willie' and weighed fourteen tons. It had a magnificent cross-country speed of 1¾ mph and never saw action, being succeeded by the familiar First World War rhomboid-shaped tanks classified 'male' or 'female' according to whether they were armed with machine-guns or six-pounder cannon. It was these which were used so effectively during the first

Across the trenches, a German A7V waddles towards a British 'Male' tank during a 1918-period battle.

large tank offensive, at Cambrai in 1917. The Germans and French were quick to follow the British lead, and in April 1918 the first tank versus tank engagement took place when three British tanks met a German A7V. The German tank won the encounter.

During the inter-war period every nation experimented with tanks apart from Germany until the rise of Hitler, for one of the conditions of the Treaty of Versailles which ended the First World War was that Germany was not allowed to build tanks, submarines or military aircraft. The pioneers of tank warfare had various ideas as to how tanks should be used in battle and gradually three main classes emerged. At the bottom of the scale were small and lightly armoured vehicles armed in the main just with machine-guns. These were the armoured cavalry, designed to be able to move quickly in the traditional reconnaissance and pursuit roles. At the other end of the scale were the infantry tanks in which speed was sacrificed for very heavy armour. They were designed to provide mobile artillery support for the infantry and to be relatively impervious to any return fire from the enemy. In the middle were the 'cruiser' tanks which were a compromise between the first two, having thinner armour than the infantry vehicles but much greater mobility, and more powerful armament than in the cavalry tanks.

Ultimately, it was the cruiser tank which was to lead to the modern generation of Main Battle Tanks (MBTs), but all tank design is a constant compromise between armour protection, armament and mobility. The cavalry tank survives in the form of various tracked reconnaissance and mechanized infantry combat vehicles but the infantry tank - which reached its peak with the multi-turreted Russian land battleships of the 1930s — has disappeared.

Different countries tried other approaches, but the concept of three different vehicles for three different tasks rather than a single versatile tank predominated. Developments were most slow in America, which had reverted to isolationism, despite the brilliance of their foremost designer, Walter Christie, whose ideas were adopted with great success by the Soviet Union. Indeed, in 1940 America did not possess a single effective tank. France wasted a great deal of time in the 'tank race' by trying to modernize old First World War designs rather than trying to develop new vehicles. By 1940 they had, however, rather surprisingly developed some of the best tanks in the world, particularly the heavily armed and armoured Char B infantry tank and the lighter Renault and Hotchkiss cavalry vehicles. Why these vehicles failed against the German Panzer divisions in 1940 was due to faulty tactical deployment rather than to any design faults, because they were in most respects superior to the German tanks.

Some of the most interesting developments took place in the Soviet Union. They suffered their share of mistakes but one concept which did gain hold was that, instead of concentrating on armour protection or speed, armament was most important. Their designers argued, correctly, that if you could eliminate an enemy tank before he could get within effective range of his own armament, then armour protection was less important. This idea resulted in the marriage of a Christie-type suspension with a powerful 76 mm gun capable of firing solid armour-piercing or high explosive shells in the T-34, the best all-round tank of the Second World War.

In Germany, following the accession of Hitler to power in 1933, the various arms manufacturers began turning out tank designs which led first to the little PzKpfw I (PzKpfw = PanzerKampfwagen, or armoured war wagon) which was essentially a training vehicle, then the PzKpfw II, III and IV which were essentially cavalry, cruiser and infantry vehicles like those of Britain and other countries. However, in many respects all these vehicles were outclassed by those of the nations against whom Hitler would shortly be warring, and their effectiveness stemmed from their organization and tactical deployment.

One of the most brilliant exponents of armoured warfare in the 1930s was a German officer called Heinz Guderian, and it was he who basically conceived the idea of blitzkrieg. In other armies, the tank was subordinated to the needs of the infantry and its deployment was conceived in essentially First World War terms. Guderian reasoned that if infantry could be made mobile by mounting them in trucks or, even better, armoured personnel carriers, so that they could keep up with the tanks, and if they were then organized into independent brigades and divisions, they would be able to punch through the enemy's front lines and wreak havoc in his rear areas, destroying the vital lines of command, communication and supply without which a modern army cannot function. And it was just these tactics which proved so effective during the German campaigns against Poland in 1939, France and the Low Countries in 1940 and, initially, Russia in 1941. Although other countries had tanks which were in many respects better than the Germans', they doled them out in 'penny packets' throughout their lines instead of concentrating them, so almost wherever they appeared, the Germans enjoyed numerical superiority.

Two views of a 1943/44 battle between German and Russian tanks using tiny 1:300 scale models. The Ju 87 'Stuka' is supported above the table on a piece of wire, one of the methods commonly used to represent aerial warfare (Darryl Reach).

Above *Another Russian-front encounter: flour has been used to simulate snow, and the Christmas cake trees look particularly effective!* (Darryl Reach).

Left *Larger-scale models look more impressive than the 1:300 scale ones, but it is virtually impossible to simulate battles larger than platoon-size skirmishes with them unless the playing area is in the garden, as here.*

One of the most successful of German tank commanders in 1940 was Erwin Rommel, and it was he who was sent out to North Africa the following year with a small force which became famous as the Afrika Korps. The strong but badly led Italian troops in Africa had been severely trounced by the British, but the Germans soon turned the scales and a see-saw battle along practically the whole coastline from the Suez Canal to Tunis developed which was to last two years. It is these Western Desert battles which are among the favourites of wargamers, for the wide open spaces and lack of towns or cities made the arena ideal for tank battles, and some of the greatest armoured clashes of the war took place on the burning sands before Montgomery managed to secure such a numerical advantage at the second battle of El Alamein that all Rommel's skill could not prevail. Relentlessly the Germans were driven back, fighting a tenacious rearguard and even scoring some significant local successes, but Anglo-American forces landed behind them in Tunisia and, since reinforcements were pitiful due to the deadly struggle in which Hitler was now locked in the east, the end of the Afrika Korps was a foregone conclusion.

In June 1941 Hitler attacked Soviet Russia using the cream of his fighting formations, and for a few short months it seemed as though his lightning campaign was going to succeed as it had done elsewhere. Hundreds of thousands of Russians were taken prisoner in great battles of encirclement, Leningrad was besieged and German troops entered the outskirts of Moscow before they were finally halted by the onset of the bitter Russian winter. The Russians, ably led by Marshal Zhukov, rallied and counter-attacked, tearing great holes in the German lines, but their advance was in turn checked and during the summer of 1942 the German Panzer divisions streamed across the seemingly endless steppes, penetrating as far east as the Caucasus Mountains. Again, however, the advance was halted by the onset of winter and the whole of the Sixth Army was trapped at Stalingrad, finally being forced to surrender in January 1943. The Soviet winter counter-offensive recovered almost all the territory captured by the Germans during the previous year, but a momentous battle between strong Russian armoured forces and the new Waffen-SS Panzer group, three divisions strong, checked the advance in March.

Both sides paused to draw breath then, in the summer, the greatest tank battle of the war took place around a small town few people had previously heard of, called Kursk. By this time new tanks

Left *Desert wargames need practically no terrain, another advantage. Here a British M3 Grant valiantly stands off a pair of PzKpfw IVs and an anti-tank gun in order to allow his command vehicle and artillery to escape* (Jeff Parker).

Below left *On the left, one of the best tanks of the war was the French Char B, seen here alongside a ruined farmhouse with a British Light Mk IV and Bren Carrier* (Jeff Parker).

Below *A Soviet KV-1 (right) emerges around a hill to intercept a German convoy in 1941* (Jeff Parker).

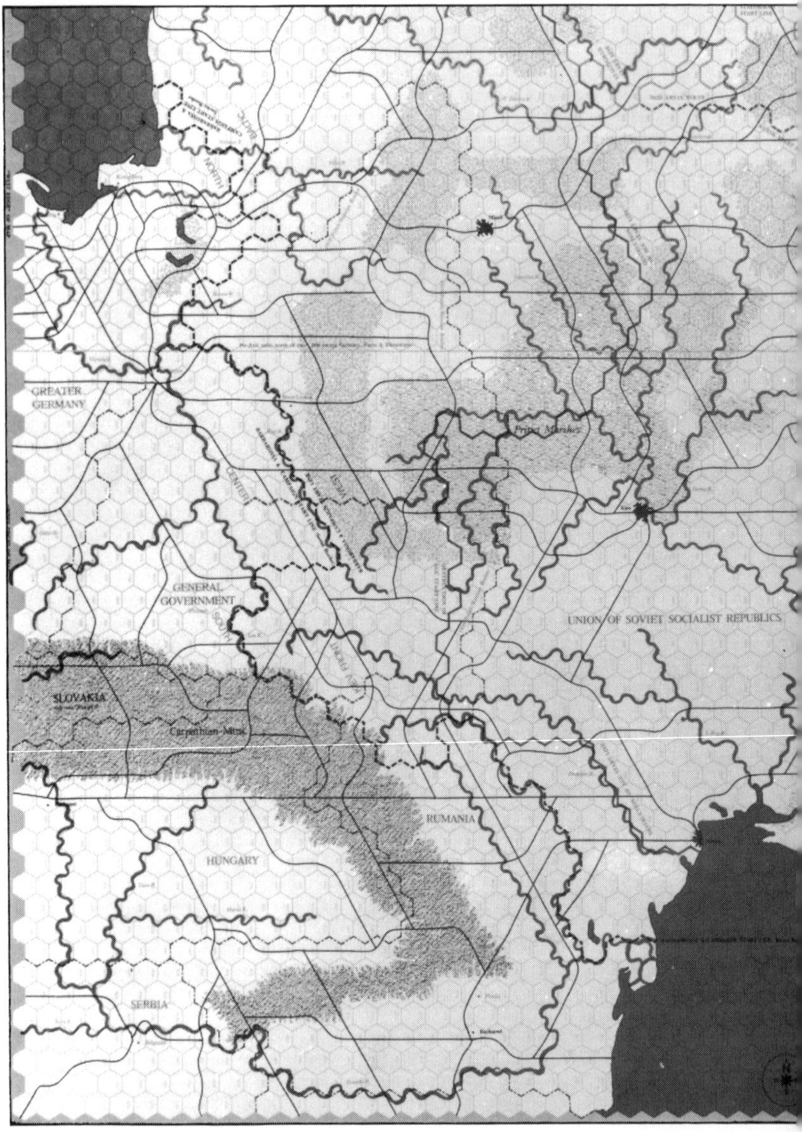

Special maps gridded in hexagons and with major terrain features pre-printed, such as this of eastern Europe and western Russia, are popular for refighting entire campaigns. When two forces, normally represented by die-

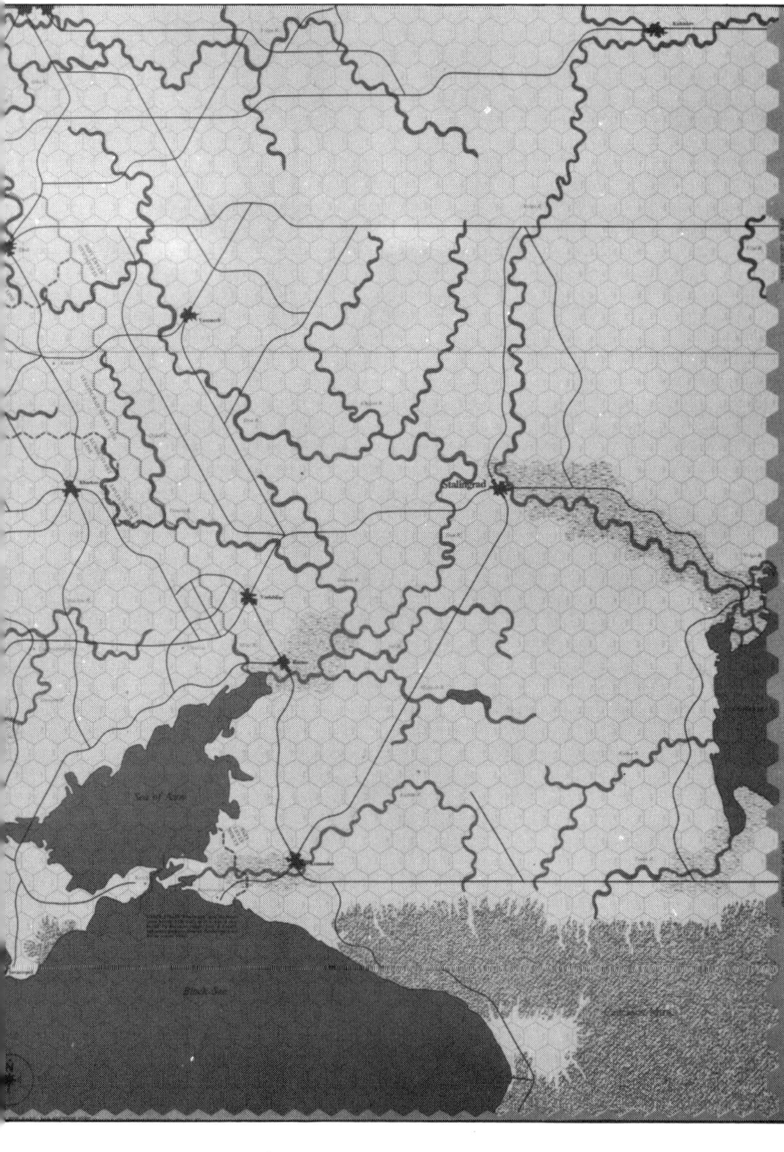

cut cardboard counters, come into contact, the result can either be determined using the playing rules supplied with such games or by transferring the action to the wargames table and using models (SPI).

Jungle warfare. British Crusader tanks and Daimler scout cars confer before crossing to the Japanese-held side of the stream (Brian Monaghan).

had come into service on both sides, notably an upgraded T-34 armed with an 85 mm gun which was nearly as effective as the infamous German '88', and the German Tiger and Panther — the latter being in many respects a copy of the T-34. The carnage at Kursk was incredible, over a thousand tanks milling about, fighting often at point-blank range. It was, as Wellington said of Waterloo, a 'close run thing' but at the end of the day the Germans gave up the attack and went over to the defensive. Kursk was, in fact, the turning point of the war for Germany and from this point onwards they were constantly on the defence on all fronts, for now the Allies began landing in Italy and divisions had to be transferred from the Russian front to meet the new threat.

Tanks were, of course, used in Italy but due to the rugged nature of the terrain over which most of the fighting took place there were

no large-scale tank battles comparable to those in Russia between British, American and German forces until after the D-Day landings in Normandy in June 1944. The Allies at this time were equipped with superior tanks to those they had possessed at the beginning of the war, notably the Sherman, an American design, and the British Churchill infantry tank and Cromwell cruiser tank. Although these vehicles were generally inferior to the German Tigers and Panthers, a new element came into play in France — Allied air superiority — and the cream of the Panzer divisions were virtually destroyed by rocket-firing Typhoons and other aircraft in the vicinity of Falaise.

The Germans were now falling back on all fronts, with the Russians pressing into Poland, Czechoslovakia and the Balkans and the other Allies streaming towards the Rhine. During the winter of 1944 Hitler tried one last gamble, assembling the best of his remaining army and SS Panzer divisions in the Ardennes — from which he had unleashed his successful offensive in 1940 — for a final counter-attack. The grandiose scheme failed, since the Panzers were able to make little headway on the narrow, icy mountain roads, and once the skies cleared the Allied fighter-bombers reappeared. The Germans continued resisting with fanaticism, although by this time there must have been an element of fatalism in it too, but the end was clearly in sight. The Allies finally succeeded in crossing the Rhine at Remagen after the tragic failure at Arnhem in the autumn of 1944, and the Russians were at the gates of Berlin. In April 1945 Hitler committed suicide and the war in Europe finally came to an end on 8 May.

Post-war, there have been many other campaigns involving the use of tanks, notably in Korea in the 1950s, Vietnam and, today, in the continuing fratricidal religious conflict between Iran and Iraq, but the most popular 'modern' campaigns among wargamers are the Arab-Israeli conflicts of 1967 and 1973, involving Israeli-modified Shermans, Centurions and Pattons and Soviet-built T-54s, '55s and '62s. As with the Western Desert campaign of 1940–43, the wide-open spaces of the Sinai Desert lend themselves to armoured warfare and there were many notable engagements, especially at the Mitla Pass in 1973.

Tank battles in miniature

Due to the cross-country speeds of tanks and other armoured fighting vehicles (AFVs), and to the range and lethality of their guns, it is difficult to fight tank battles as wargames using the same scales as before. For this reason most wargamers use models in 1:300 scale (ie, one inch to 25 feet). These tiny tanks, the size of a thumbnail, are manufactured by a variety of firms in incredible detail, many even having revolving turrets, and some wargamers manage to paint national and even divisional insignia on them — although others are content to just give them a coat of green, yellow or grey paint. Because of their size, these models are often sold in packs of two or more, and buying some of these is by far the cheapest way to get into wargaming. Model houses and other terrain features can also be purchased in this scale, although it is easy to make them yourself or to improvise — I find the type of Christmas tree sold as cake decorations ideal, for example, but be prepared for raised eyebrows when you go into a shop and ask for fifty!

Playing rules for armoured warfare can become incredibly

complex and those for combined operations involving AFVs, infantry, aircraft and perhaps even landing craft even more so, and a fully-detailed set would fill an entire book this length so there is only room here for a very simple, but nevertheless playable and enjoyable set. (Patrick Stephens did, in fact, publish a set of five books called *Tank Battles in Miniature* during the 1970s which covered the subject in great depth and, although these are now out of print, your local library should be able to obtain copies.)

All movement and firing is simultaneous as before, but to reflect the greater speed of movement and firing, and the increased weapons ranges, different ground and time scales are needed. If we halve the game turn and make it thirty seconds, and make one inch on the table represent fifty yards, it works out very well: a tank moving at an average of 25 mph will thus cover $\frac{(25 \times 1,760)}{120} \div 50$ inches per turn, or just over seven inches. Similarly, a gun with an actual range of 2,000 yards will have a scale range of forty inches, so play on a normal-sized table is perfectly feasible.

It is impossible here to take into account the enormous differences in armour thicknesses on tanks and other AFVs, which during the Second World War could range from 5 mm — adequate to protect the occupants against small-arms fire only — to 185 mm in the case of the frontal armour on a Tiger II or King Tiger. Similarly, the armour on the sides and rear of a tank was normally thinner than that on the front. Fortunately, armour thickness is usually proportional to the size of gun forming a tank's main armament so for wargaming purposes we can merely classify all tanks as 'light', 'medium' or 'heavy'. The most common types are thus classed as follows (size of main armament after each name in brackets):

France Heavy — Char B1 (75 mm *and* 47 mm); Medium — Char D2 (47 mm); Somua S-35 (47 mm); Light — Hotchkiss H-35/H-39 (37 mm); Renault R-35 (37 mm).

Germany Heavy — Tiger II (88 mm); Tiger I (88 mm); Medium — PzKpfw V Panther (75 mm); PzKpfw IV (75 mm); PzKpfw III (50 mm); Light — PzKpfw 38(t) (37 mm); PzKpfw II (20 mm).

Italy Medium — M13/40 and M14/41 (47 mm); Light — CV33 (13 mm).

Japan Medium — Type 97 (57 mm); Light — Type 95 (37 mm).

Soviet Union Heavy — JS-3 (122 mm); KV-II (152 mm); KV-I (76 mm to 1943, thereafter 85 mm); Medium — T-34 (76 mm to 1943, thereafter 85 mm); Light — BT-7 (76 mm); T-26 (37 mm); T-60 (20 mm).

United Kingdom Heavy — Churchill (75 mm); Matilda II (40 mm); Valentine (57 mm); Medium — Sherman (75 mm); Sherman Firefly (76 mm); Grant (75 mm *and* 37 mm); Cavalier (57 mm); Centaur (57 mm); Covenanter (40 mm); Cromwell (75 mm); Cruiser Tanks Mks 1-4 (40 mm); Crusader III (57 mm); Light — Vickers Light Tanks Mks 1-6 (machine-guns only); Tetrarch (40 mm).

United States of America Heavy — M26 Pershing (90 mm); Medium — M4 Sherman (75 mm); M3 Lee (75 mm *and* 37 mm); Light — M3/M5 Stuart (37 mm); M24 Chaffee (75 mm).

It should be noted that many tanks, particularly British ones, were progressively up-gunned as the war progressed. The Churchill, for example, started life with a 40 mm 2pdr, then had a 57 mm 6pdr and finally a 75 mm gun or 95 mm howitzer, but in the above figures only the most predominant weapons are featured.

For wargames purposes, we classify guns as follows: weapons over 80 mm can penetrate the armour of light tanks at up to 2,000 yards (40 inches), of medium tanks at up to 1,200 yards (24 inches) and of heavy tanks at up to 500 yards (10 inches). Weapons of between 57 mm and 76 mm can penetrate the armour of light tanks at up to 1,600 yards (32 inches), of medium tanks at up to 900 yards (18 inches) and of heavy tanks at only 300 yards (6 inches). Weapons of 37 mm to 47 mm can penetrate the armour of light tanks at up to 1,000 yards (20 inches), of medium tanks at up to 400 yards (8 inches) and of heavy tanks *in the sides or rear only* at 100 yards (2 inches). Weapons below 37 mm in calibre can only penetrate the armour of light tanks at up to 600 yards (12 inches) or of medium tanks in the sides or rear at 100 yards (2 inches).

Tanks may fire at any time but suffer a penalty for reduced accuracy if firing on the move. Movement rates are as follows: heavy tanks four inches per turn, medium eight and light twelve. To engage a target, first indicate to your opponent which tank is firing at which target, and *then* measure the range. (This prevents you changing your mind and choosing another target if you discover that your chosen target is out of range or that your own tank's gun cannot harm it at the measured range!) Having measured the range and determined that you *can* damage the target, the next procedure is to throw a dice to see whether you have actually hit it. Throw one ordinary dice. At long range you need a five or a six to secure a hit, at medium range a three, four, five or six while at close range anything except a one denotes a hit. However, from your dice score deduct -1 if the target is moving or behind cover, and an additional -2 if your own tank is moving. (The long, medium and short range bands are as above, thus for an 88 mm gun long range is 25 to 40 inches, medium 11 to 24 inches and short anything up to 10 inches.)

A major difference between these and the earlier rules is that a tank is not necessarily 'killed' once a hit has been achieved, so once you have scored a 'hit' it is necessary to assess the damage. You may just have inflicted minor damage or briefly concussed the driver, so a second dice throw is now required. A one means the target is immobilized for one turn but may still fire; a two that it is immobilized for three turns and cannot fire during the first of those; a three means that it has suffered major damage immobilizing it for the rest of the game, but its gun can still fire; a four means it may still move but cannot fire during the next turn; a five that it can move but not fire for three turns; and a six destroys it completely. If you hit the side or rear of a target with a gun of greater than 50 mm calibre, add +1 to your score.

The above factors are accumulative so that, for example, if you score a two against a target in one turn, and then score a one against it in the next, the effect is as if you had scored a three in the first place. A further hit scoring one in the subsequent turn would not, however, miraculously restore the target's mobility, but *would* prevent it firing back. To keep track of the status of each tank it is advisable to number them beforehand, either by means of painting individual numbers on them or by applying small self-adhesive stickers, and listing these numbers in a notepad against which you can jot the appropriate damage factors.

Armoured battles can, and usually did, of course, involve infantry as well, and if you wish to introduce these the simplest method is to assume that each section is equipped with a bazooka, Panzerfaust or similar weapon, and treat them as 'light tanks' in all except their movement factor which should be one inch per turn. If you have motorized infantry in trucks or half-tracks, they should move at the same speed as medium tanks but must remain stationary for a full turn in order to allow the troops to 'debus' and deploy before they can then fire. As you can see, such refinements can continue indefinitely and one of the principal challenges you will encounter as a wargamer is that of further and more detailed research in technical books on the weaponry of your chosen period, on the armies' appearances and so on. Fortunately, these days there are literally hundreds of reference books available compared with the meagre few when I first started wargaming!

NAVAL WARFARE

Although there are no apocryphal tales of Alexander or Nelson using model ships to plan their battles, it is logical to assume that some naval commanders down the ages have used wargames in some form or another while plotting their future actions. The naval wargame in its present form, though, only goes back as far as the turn of the 20th century and a naval kriegspiel invented by Fred T. Jane, the legendary founder editor of *Jane's Fighting Ships*, which was sold in boxed sets for three guineas — a sizeable sum of money in those days. This game was principally intended for naval officers and the general public did not really become interested in naval wargaming until the American writer Fletcher Platt devised a set of rules in the 1930s so simple that any beginner could understand them. As happened with H.G. Wells' *Little Wars* earlier, though, a global conflict killed off interest in this until the 1960s when wargaming as a whole began to become popular.

One of the problems in naval wargaming is that it is very difficult to restage an historical battle such as Trafalgar or Jutland because, even if you select identical conditions of sea and wind, few wargamers will use identical tactics so the battles might as well be entirely fictitious. In a land battle, tactics are largely dictated by terrain so a refight of Waterloo will be instantly recognizable but at sea this is not the case and the circumstances leading up to a naval battle and the reasons *why* it took place become more important than the when, where or how. I have not mentioned campaigns earlier, but in a naval context in particular it is far more satisfying to refight an entire campaign, or the critical part of one, than to merely say 'right, my fleet's sailing in from this side of the table, yours from that, and may the best man win'. In a campaign, fought with a regular group of opponents or even by post, you can introduce all the elements of planning, logistics, lines of communication and supply and even politics (because you need a reason for a war). Thus, when your armies or fleets of ships do come together in battle, there is a reason for it and rather than fighting to the bitter end, one side or the other may decide that it is better in their long-term interests to allow discretion to become the better part of valour and voluntarily retire from an engagement — if they can, of course.

Like war on the land, naval warfare can conveniently be divided into four broad periods: ancient galley warfare, the age of sail, the age of steam and the modern period from the Second World War onwards with oil-fired and even nuclear vessels.

Ancient galley warfare

In the very early days, warships were not designed so much as weapons in themselves but just as means of carrying soldiers to do battle with the enemy's soldiers, so tactics were simple. Ships would grapple together and, while the oarsmen crossed their fingers, the marines aboard each would leap into hand-to-hand combat. Archers would, of course, fire at the enemy during the approach to contact, and some ships were equipped with large crossbows or stone-firing ballistae. The Romans even fitted their ships with towers from which archers could continue to fire into the enemy even after two ships were locked together.

It was the development of the ram which revolutionized early naval warfare, making the galleys themselves potent weapons capable of sinking other vessels by punching holes in their hulls below the waterline. Whereas, hitherto, 'tactics' had consisted of two opposing fleets simply lining up facing each other and

Ancient galleys locked in mortal combat with boarding ramps lowered and archers in the 'towers'. These home-made models by Paul Hague are ideal for use with 25 mm figures (Brian Monaghan).

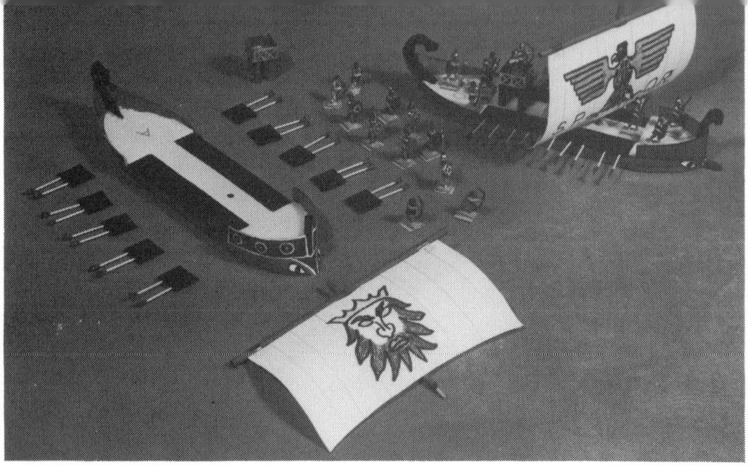

Component parts of home-made Roman galleys, showing how they can be taken apart to depict battle damage (models by Paul Hague, photography Brian Monaghan).

ploughing straight into contact, the development of the ram necessitated more thought and two basic tactics were evolved known as the *diekplous* and *periplous*. In the former great co-ordination between individual ships was required, for the idea was that one vessel would deliberately try to swipe the oars from one side of an enemy vessel, leaving a vulnerable flank into which a second ship could plunge its ram. The *periplous* involved trying to outflank one side or the other of the enemy fleet and could normally only succeed with the help of numerical superiority, because the enemy's front obviously had to be 'pinned' while the outflanking manoeuvre was taking place.

The main motive power of all warships up to, really, the Renaissance was the oar, since the clumsy square or rectangular sails with which ships were fitted could only run more or less with the wind, and vessels could not tack with any degree of ease or sail close to the wind. Ships averaged 100 to 150 feet in length and were normally crewed by two or three banks of oarsmen, such vessels being known respectively as biremes or triremes. Records indicate the existence even of ships powered by five banks of oarsmen (quinqueremes), but nobody has yet worked out satisfactorily how they were seated! Model galleys of various types are available from several manufacturers in a variety of scales, and you can even buy 'waterline' models (ie, with flat bottoms) in plastic kit form in 1:72 scale, meaning that you can use ordinary 25 mm model soldiers for their crews and complements of marines.

The age of sail

During the 13th and 14th centuries great advances were made in ship design. The lateen, or triangular sail — which permits vessels to sail closer to the wind — came gradually into use, different methods of construction were employed and ships started to feature more than one mast. Cannon were mounted in warships from early in the 14th century, three English vessels being

Napoleonic naval warfare — one fleet running, the other close hauled, the French and British approach each other almost head-on (models by Paul Hague, photography Brian Monaghan).

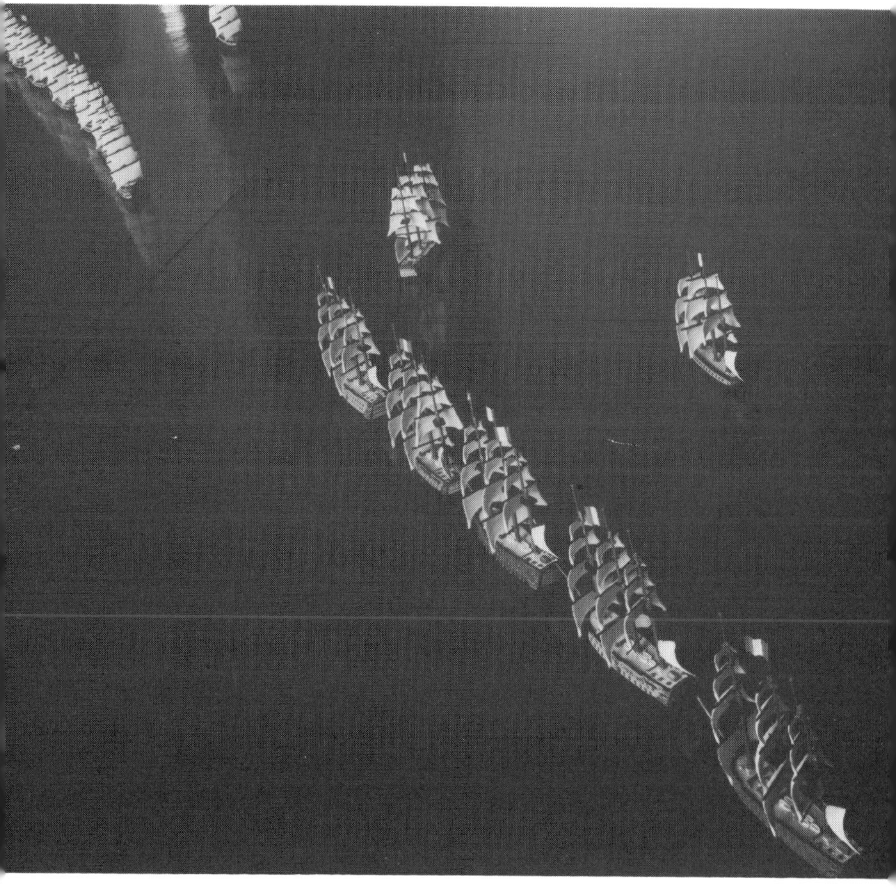

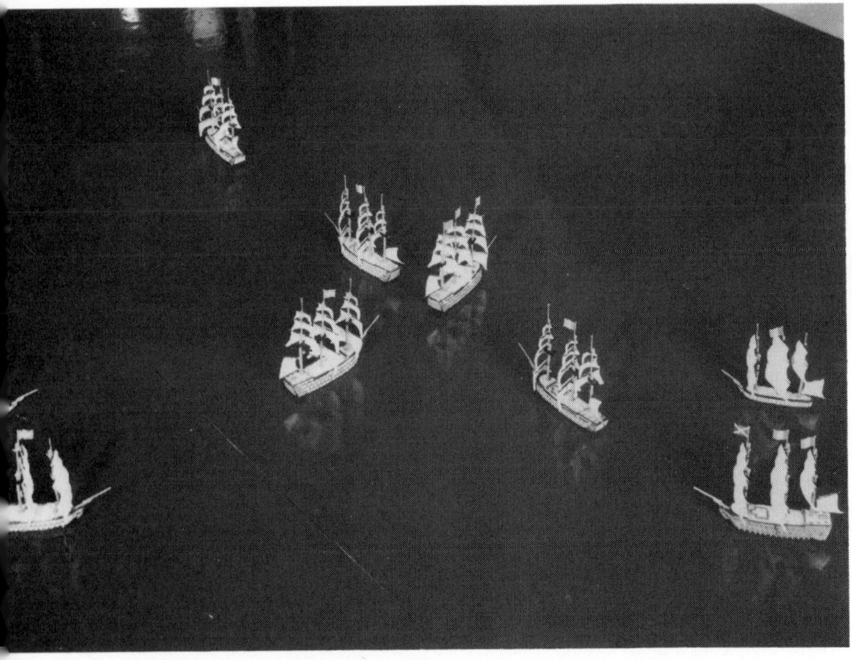

A different tactic. Here the British fleet has penetrated the French line at right angles, allowing full raking broadsides (models by Paul Hague, photography Brian Monaghan).

reportedly so equipped in 1338 while the Spanish used them in action in 1372. The number of sails carried gradually increased and the Roman invention of the bowsprit was rediscovered, permitting far greater flexibility and speed of manoeuvre. A stern-mounted rudder and tiller replaced the earlier steering oars and ships gradually became larger, often with fore and aft 'castles' which provided both accommodation space and firing platforms. However, it was soon found that these ruined a ship's centre of gravity so guns began to be mounted in the hull, below the main decks, with firing ports covered with shutters to keep out heavy seas. These were invented by Descharges of Brest in 1500, and it was with such ships that Drake was able to defeat the Spanish Armada in 1588, for the English relied upon firepower as their principal weapon while the Spaniards adhered to time-honoured boarding tactics.

The fleets are now well and truly engaged, several boarding actions are taking place and some masts and rigging have been lost (models by Paul Hague, photography Brian Monaghan).

Designs were constantly refined and improved throughout the Elizabethan and Tudor eras, and vessels emerged with four masts — the mainmast and foremast being square-rigged and the stern pair, or mizzens, lateen-rigged. A lateen-type sail called a jib was also rigged between the foremast and the bowsprit which permitted ships to tack, or change direction against the wind, much more easily. Rigs were constantly thus improved throughout the 17th and 18th centuries, better techniques for 'staying' masts enabling them to become taller and carry a greater area of canvas. More scientific approaches to the hitherto almost empiric science of ship design resulted in two main classes of warship gradually evolving in the period leading up to the Napoleonic Wars, the heavily gunned 'man o' war' and the lighter, faster and more manoeuvrable frigate. All warships by this time were broadly classified by the number of guns they carried, men o' war generally having 72 to 80 and frigates 24 to 32.

The introduction of broadside-firing cannon aboard warships had a notable effect upon tactics, in that the former line-abreast naval formations became progressively line-astern, permitting each ship in a fleet to bring maximum firepower to bear as it passed alongside or through the enemy formation. The spacing between each ship in such a line thus becomes crucial: if two 'fleets' of six ships engage each other in line, one with its vessels 100 yards apart and the other 200, the fleet with the smaller intervals will be able to concentrate greater firepower than the enemy can return. The fact that the two formations are evenly matched in numbers of ships does not matter — it is the number of guns which can be brought to bear which is decisive. By the same token, if an admiral could bring two lines of ships either side of an enemy line, he could concentrate double the firepower against each vessel, or even more if his own ships were in a tighter formation. Crossing the enemy's line in its centre, enabling each friendly ship to bring full broadsides to bear against the ships either side of the break, and then turning with the wind to run down half of the enemy flank with a whole fleet was even more effective when it could be practised, and it was such a tactic which brought English success at Trafalgar.

The key element in 'the age of sail' was obviously the wind, and the fleet which managed to manoeuvre so as to have the prevailing wind with it was at an obvious advantage. However, a weaker force would normally endeavour to manoeuvre to leeward of a stronger one so that it would have the advantage if forced to flee. Similarly, while a strong force would seek to hole its opponents below the waterline, a weaker one would use chain shot (lengths of chain fired from the guns) or red-hot shot to try and break or set fire to the enemy ships' rigging and thereby prevent the enemy pursuing them if they were forced to retreat. All this changed after the Napoleonic Wars with the introduction of steamships.

Model ships suitable for wargaming the 'age of sail' are plentiful, ranging from tiny white metal castings in 1:1,200 scale through to elaborate (and expensive!) plastic kits which could be cut down into waterline models. All products of this type are advertised regularly in magazines such as *Military Modelling.*

The age of steam

The first coal-fired, steam-propelled vessel to operate commercially in Europe was Henry Bell's *Comet* on the River Clyde in 1813, but for many years ocean-going steamships (still wooden-hulled) continued to place more reliance on their sails, using the engine mainly as an auxiliary source of power for manoeuvring in harbour or when becalmed. Early engines were inefficient and uneconomical to run, although the P & O liner *Himalaya* was purchased by the Royal Navy and operated very successfully as a troop transport during the Crimean War. The development of the compound engine in the 1860s and then the triple-expansion engine in the 1880s improved matters considerably, and gradually the swift and graceful clipper ships lost their supremacy on the world's trade routes.

'Ironclad' warships were first used during the American Civil War.

The age of steam — ironclads, their sails furled, steam into battle in a fictitious late 19th century encounter (models by Paul Hague, photography Brian Monaghan).

To begin with such vessels were essentially men o' war built wholly or partially of iron, relying on sails as well as steam propulsion and having their guns arranged in broadside batteries behind ports as in the Napoleonic period. Gradually, however, turrets or barbettes (open-topped turrets, the guns firing over their walls) came into use, and by the late 1880s warships were beginning to resemble in outline those we know today. The same improvements in artillery which were taking place on the land also took place at sea, rifled breech-loading cannon replacing the old muzzle-loading smooth-bores. Strangely, the ram was also reintroduced for a short while, alongside new naval weapons such as torpedoes and mines.

By the turn of the century 'ironclads' were giving way to the new 'dreadnoughts' (named after the first ship to be thus christened, HMS *Dreadnought* in 1906), and it was with these that the navies of Britain, France and Germany went to war in 1914. After the unification of Germany in the 1860s, both Germany and Britain had become engaged in a bitter struggle for colonial supremacy, particularly in Africa, and Germany — never previously a real seagoing nation — entered into a naval arms race to try and match the global supremacy of the Royal Navy.

Alongside the dreadnoughts, other classes of warships — battle-cruisers, cruisers, destroyers and torpedo boats — came into being, all built of steel and all propelled by triple-expansion steam engines. The battleship was queen of the seas, armed with an average of ten twelve-inch guns capable of firing with accuracy up to 20,000 yards, while the cruisers and destroyers acted in the reconnaissance role, tracking down the enemy and, hopefully, pinning him until a friendly force of battleships could arrive on the scene. They also acted as patrol vessels and convoy escorts, but their lighter armour and armament meant they could not stand up to dreadnoughts in a straight fight. However, a new and sinister weapon was already in existence which, coupled with naval air power, would ultimately render the battleship obsolete. This was, of course, the submarine.

Submarines were not a new invention, having been used as early as the American War of Independence in 1776 and an American, Robert Fulton, demonstrated his own design of such a boat to the French navy during the Napoleonic Wars, but it was not taken up. Ironically, therefore, credit for the first 'proper' submarine (earlier ones all being hand-propelled by means of oars (!) or screws) belongs to a Frenchman, C.D. Goubet, who in 1881 designed a small boat powered by electricity. By the time the First World War began in 1914 most navies had submarines in their fleets, powered by petrol engines while running on the surface and by electrical batteries when submerged. The Royal Navy had 96 of these vessels at the outbreak of war, capable of up to ten knots submerged and armed with either two or three torpedo tubes. Development might have been even faster had not an international agreement effectively prohibited submarine use against merchant shipping by stating that such vessels must not be sunk without warning and that their crews must be allowed to take to their boats within reach of land. Germany, a signatory to this agreement, discarded it during the First World War, however, and ultimately sunk nearly sixteen million tons of Allied and neutral merchant shipping, almost bringing Britain to her knees in 1917.

There are few models available for the larger part of the 'age of steam' apart from ironclads and 'monitors' for the American Civil War, and even when one gets into the dreadnought period most of those which are sold commercially are tiny 1:1,200 or 1:2,400 scale white metal miniatures. Although one of the most fascinating and innovative periods in naval history, the age of steam is ill-favoured amongst wargamers in general, and even such classic encounters as Jutland or Coronel are rarely encountered on the wargames table. Far more popular is the Second World War.

The modern period

As always, war is the 'mother of invention' and the First World War was no exception in this respect. By the time the Second World War began, although warship classes were still broadly the same they were all oil-fired, giving greater speed and endurance, better armoured and more efficiently armed not just with larger guns in many cases but with far superior fire control directors. Other changes included provision on all vessels for small-calibre anti-aircraft weapons, whose number increased drastically as the war progressed, plus the gradual introduction of sonar, to detect submarines, and radar to detect surface units and aircraft. It was naval aviation which really changed things the most radically, though.

The value of air power at sea had been recognized as early as 1908, and in 1911 the first aeroplane was successfully launched from a ship, the British battleship *Africa*. During the First World War various merchant ships were converted so as to be able to fly off aircraft. These were all seaplanes, however, of limited value except

in the reconnaissance role, and it was not until 1917 that a conventional aircraft achieved a successful landing on a naval vessel, the converted British heavy cruiser *Furious*. The world's first true aircraft carrier with a real flight deck, however, was the Italian liner *Conte Rosso* which was commissioned into the Royal Navy as HMS *Argus*. Post-First World War treaty arrangements prohibited much naval construction, particularly by limiting the size of tonnage, and it was not until 1935 — after the rise of Hitler to power in Germany — that serious efforts were made to construct new carriers. (Ironically, Germany never commissioned any carriers.) These, both the large fleet carriers capable of carrying approximately sixty fighters, fighter-bombers and torpedo-bombers, and the smaller escort carriers which were converted from merchantmen to help defend convoys from the U-boat menace, could justifiably be said to be the weapons which, more than any other, won the war — particularly that in the Pacific and Indian Oceans against the Japanese. After the Second World War it was apparent that the aircraft carrier had supplanted the battleship as the queen of the seas and most navies were quick to scrap or 'mothball' their obsolete monsters — although the tide may be turning again for the United States Navy has in recent years re-activated the battleships *New Jersey* and *Iowa* with both missiles and sophisticated

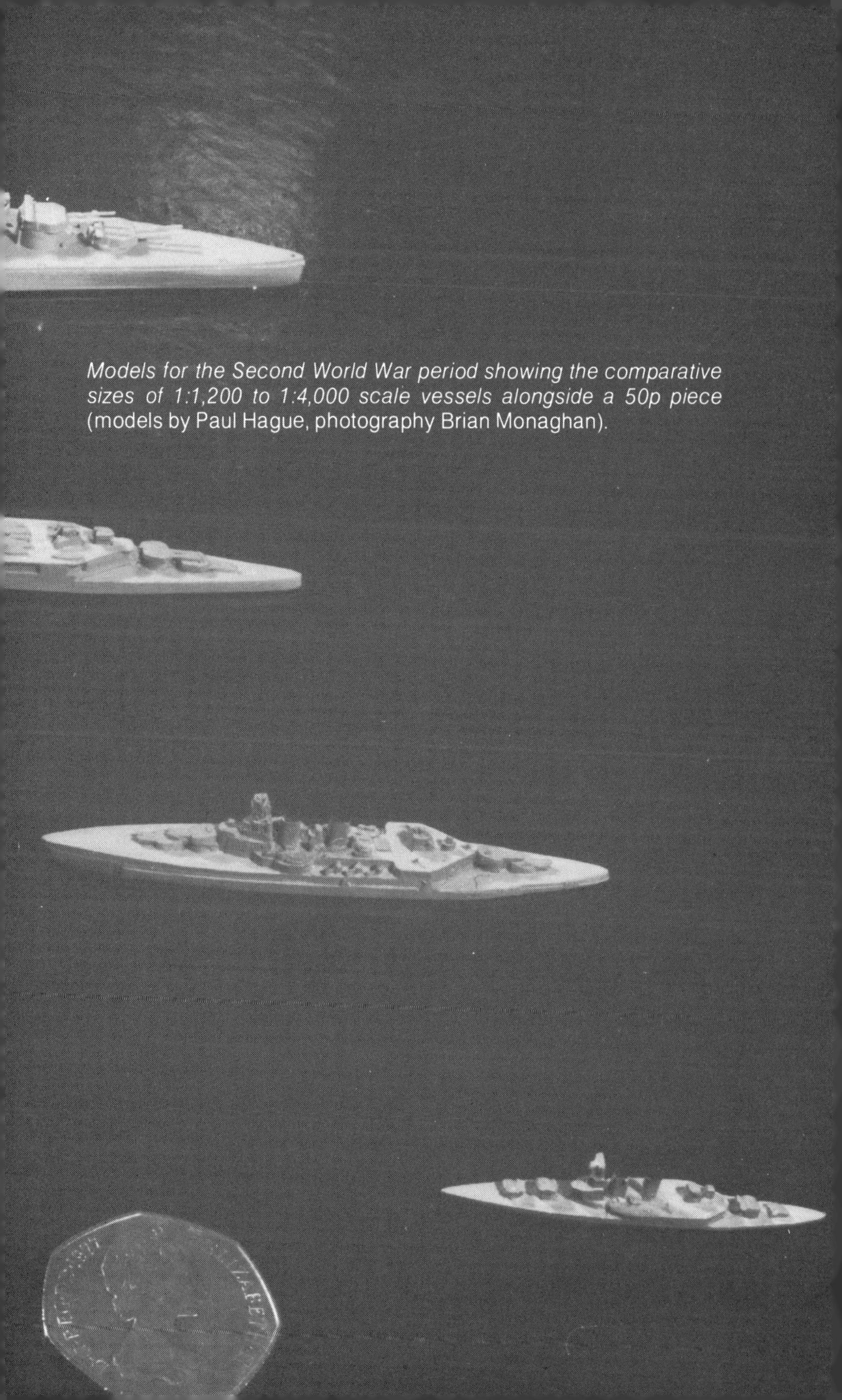

Models for the Second World War period showing the comparative sizes of 1:1,200 to 1:4,000 scale vessels alongside a 50p piece (models by Paul Hague, photography Brian Monaghan).

communications gear to enable them to act as command ships, and has plans to do the same with *Missouri* and *Wisconsin*. Carriers, though, continue to be the principal element in most navies, alongside nuclear hunter-killer and missile submarines. The Falklands' conflict of 1982 showed the Royal Navy how unwise it had been to have scrapped its last conventional carrier, *Ark Royal*, in favour of the new 'through-deck cruisers', despite the huge success of the Harriers and Sea Harriers from *Hermes* and *Invincible* during that campaign.

During the Second World War there were few major naval engagements between 'traditional' capital ships, the battle of the River Plate and the chase of the German battleship *Bismarck* being the two most notable exceptions. Otherwise, most capital ships succumbed to air power — Pearl Harbor and the sinking of the *Prince of Wales* and *Repulse* springing to mind as immediate examples. The submarine, too, proved vulnerable to air power and, despite the invention of the schnorkel which permitted U-boats to recharge their batteries and air supplies while still submerged, approximately ninety per cent of German U-boat crews met their deaths at sea by the end of the war. Modern submarines, in particular nuclear ones, enjoy greater security, being able to operate at great depths and speeds for indefinite periods of time, making their location extremely difficult despite advances such as helicopter-borne 'dunking sonar' and surveillance from orbital satellites equipped with sensitive cameras and scanners able to detect minute variations in sea colour or temperature.

Apart from providing close-range anti-aircraft coverage and fire support for ground forces, the gun has also disappeared as the warship's principal weapon, to be replaced by anti-submarine, anti-ship and anti-aircraft missiles of far greater range, accuracy and lethality — witness the effectiveness of the French-built Exocet missile in the Falklands' conflict. All of this makes modern naval warfare — which can only be an integrated air/sea exercise anyway — far too complicated to be fought on any wargames table without the help of a reasonably powerful computer, and puts it outside the scope of this book. The rules which follow are, therefore, limited to pre-1945.

Sea battles in miniature

Most people have played 'Battleships' at some time in their lives, but having read this far you will realize that this is as far removed from a real wargame as is Ludo, say. Moreover, in naval warfare there are often more factors which have to be taken into account than in land warfare, so even a 'simple' set of playing rules has to be moderately complicated. Movement is straightforward in the sense that it is easy to lay down a ground scale — or rather, sea scale — for the wargames table and relate a vessel's known actual speed to this, but movement rates — particularly prior to the age of steam — will be influenced by the force of the wind and state of the sea, while sailing ships will also be able to move faster while running with the wind than when tacking into it. The radius of a vessel's turn, which is related to both a ship's length and the speed at which it is moving, must also be taken into account. Similarly, the number of instances when a ship is sunk by a single salvo, as in the case of the British battlecruiser *Hood* in the Second World War, are extremely rare, so we need a damage assessment system. However, the basic elements of play remain essentially the same as for land warfare.

1 All movement is simultaneous.

2 Each model ship represents one actual vessel.

3 Each game turn, or move, represents one minute of actual combat time.

4 For the purposes of simplicity, it is assumed that all model ships are to 1:1,200 scale, and the sea scale is proportional to this (ie, one inch on the table represents a hundred feet). All movement is measured in knots, a speed of ten knots being ten nautical miles per hour. Since a nautical mile is approximately equivalent to 2,025 yards, speed can be directly proportional to ground scale, thus in one minute a vessel travelling at one knot will cover 33¾ yards (2,025 ÷ 60), or a fraction over a hundred feet. We can thus say that a vessel moving at ten knots will cover ten inches per turn, and so on.

5 Apart from dice and a measuring tape, the other equipment you need to construct is 'turning circles', and unfortunately you need to make these from one inch diameter up to the equivalent in diameter of your fastest ship's maximum speed. Thus a galley travelling at ten knots would sail round a quarter of a ten-inch circle in order to turn through 90°, or half-way round the same circle in order to reverse its course. Fortunately, steamships and more modern vessels *are* able to turn more nimbly than sailing ships, otherwise we would need turning circles up to forty or even more inches in diameter. Ironclads powered by steam and sail use turning circles proportionate to two-thirds of their speed (thus a ship travelling at eighteen knots would use a twelve-inch circle), while ships fully powered by steam or diesel engines use circles proportionate to half their speed. These turning circles are absolutely vital, in particular when manoeuvring to ram or board.

6 As before, although movement and firing are deemed to be simultaneous, we measure it in two phases, long-range at the beginning of each turn and short range immediately before the phase of hand-to-hand combat after two ships have locked together. (Short-range fire and mêlées become progressively rarer in the later historical periods but the same methods can be used if recreating, for example, the storming of the German prisoner-of-war vessel *Altmark* by men from the destroyer *Cossack* in the Second World War.)

7 The morale of a ship's captain and crew must be tested after each round of mêlée during a boarding action and the rules for land warfare may be used. In other circumstances, a ship's morale is tested after it has suffered a certain degree of damage to its hull, guns or rigging, as discussed later.

8 A record card or simple list on a sheet of paper is needed to keep track of damage as well as for the writing down of orders, which *does* need to be done at the beginning of each turn in naval wargames.

9 Wind direction is important at sea, so at the beginning of the game nominate one side (or corner) of the table as north, then throw a single ordinary dice. One means the wind is from the north, two from the south, three from the east and four from the west. If a five or six are thrown, throw again until you achieve a score of between one and four.

10 Wind strength is also important and is resolved in the same way, one denoting a light wind, two a moderate wind, three a strong wind and four a gale. Galleys caught in a gale will not survive long so unless the battle being fought is part of a campaign — in which case the ships are unlikely to be at sea anyway — I suggest that in games with galleys you ignore a dice throw of four and instead have one and two for a light wind, three and four for moderate and five and six for strong.

Movement for sailing ships is dictated entirely by wind speed and direction. Galleys can use the wind to help their oarsmen when it is with them but must furl their sails or even — as often happened — step their masts before rowing into it. Steamships are only really affected by strong winds or gales, and then relatively marginally. However, it is obvious that all vessels can move more quickly with a following wind, less so when it is coming from the side and least of all when it is from dead ahead. The first two courses are calleo 'running' and 'reaching' respectively. A sailing vessel only comes head-up into the wind when it is 'going about' or, in other words, using the force of the wind to turn the bows so that the wind hits the sails from the other side of the ship and it continues its voyage on a course roughly 90° to that it was formerly pursuing. This process is called 'tacking' and if you draw a straight line on a piece of paper with an arrow in one direction denoting the wind direction, then draw a continuous zig-zag line bisecting this at 45° angles throughout its length, you will see the meaning clearly. At each 'reach' the ship is sailing with the wind from either port or starboard, making slow diagonal progress into the wind.

A steamship can, of course, head straight into the wind although the fact that the seas will be piling up in front of it in waves of increasing height the longer the wind lasts will hinder its progress. This is particularly true of aircraft carriers. Nor, in the case of pure sailing ships, does the strength of the wind automatically confer higher speed because, above a certain level, sails have to be furled (taken down) or reefed (wound around their booms or spars and tied down so as to present a smaller sail area to the wind) unless they, their spars and even the masts be carried away.

Thus, speed is dependent upon both wind speed and direction and a vessel running before a gale would, in practice, normally cover a smaller mileage per hour or minute than one tacking into a moderate breeze. In a gale, with its attendant heavy seas, firing guns was either impossible or impossibly accurate until the relatively modern inventions of computerized, inertial navigation, radar-controlled tracking and guidance systems. The following table therefore gives movement rates per turn for various classes of warship dependent upon wind strength and direction. 'Bows' means the wind is blowing within 45° either side of the vessel's bow, 'sides' within 45° of either side and 'stern' within 45° either side of the stern. Pure sailing vessels cannot sail 'bows on' but turn into the wind so as to tack, so forward movement is effectively zero and the figures given are for tacking rather than sailing a straight line course.

Firing

All firing described here is ship-to-ship. If you wish to introduce air-craft into naval games, the following chapter will provide some ideas. Aircraft cannot, however, be launched or landed on carriers under strong wind or gale conditions. Remember, too, that carriers have to

Movement rates in knots (inches per turn)

Ship type	Light wind			Moderate wind			Strong wind			Gale		
	Stern	Sides	Bow	Stern	Sides	Bow	Stern	Sides	Bow	Stern	Sides	Bow
Galleys												
Sail only	4	3	0	5	4	0	6	4	0	—	—	—
Oars only	5	4	3	6	4	2	7	3	1	—	—	—
Oars & sail	6	5	1	8	5	0	5	3	0	—	—	—
Sailing ships												
Over 80 guns	8	6	3	10	8	4	12	10	5	3	2	1
Over 40 guns	10	8	4	12	10	5	14	12	6	6	4	2
Below 40 guns	12	10	5	14	12	6	16	14	7	8	6	3
Ironclads												
Sail only	8	6	3	12	10	5	10	8	4	3	2	1
Steam only	16	14	12	16	14	10	14	12	8	8	6	4
Sail & steam	14	12	10	16	14	8	12	10	5	8	6	2
Battleships												
World War 1	21	20	19	21	20	18	20	19	17	15	12	10
World War 2	32	31	30	32	31	29	30	28	26	25	20	15
Battlecruisers												

World War 1	23	22	21	23	22	20	17	15	13	15	13	11
World War 2	32	31	30	32	31	29	30	28	26	25	20	15
Light cruisers												
World War 1	27	26	25	27	26	24	25	23	21	20	15	10
World War 2	33	32	31	33	32	30	30	28	26	22	16	12
Destroyers												
World War 1	37	36	35	37	36	34	33	30	26	24	18	14
World War 2	40	39	38	40	39	37	34	31	27	26	20	16
Frigates												
World War 2 only	30	29	28	30	29	27	25	23	20	20	15	10
Submarines												
World War 1	15/10*	14	13	15	14	12	10	8	6	—	—	—
World War 2	18/12*	17	16	18	17	15	12	10	8	—	—	—
Fleet Carriers												
World War 2 only	31	30	29	31	29	27	25	22	20	15	12	8
Escort carriers												
World War 2 only	18	17	16	18	16	14	15	13	11	10	8	6

*The second figure in each case is the maximum submerged speed which is essentially unaffected by surface conditions.

Ships do not, of course, charge around the ocean at full speed all the time and would normally cruise at around half maximum but may increase or decrease their speed by 25 per cent of maximum each turn.

turn into the wind to launch or land aircraft. Because of the enormous variety of armament which could be listed, I have 'bracketed' weapons so that, for example, a Second World War ship with fourteen-, fifteen- or sixteen-inch guns will have the same firepower.

Bows, crossbows and muskets, etc, have the same ranges as in the rules for land warfare, ie, approximately a hundred yards or three inches to our naval wargame scale, and both firing and mêléeing take place using the land warfare rules. To do this, you can actually use model sailors and marines, or more simply list against the name of each vessel its crew complement and then note casualties as they occur. The same *could* be done for the effects of artillery fire, but it becomes terribly unwieldy and time-consuming and it is more usual, therefore, to score 'damage points' against various parts of the target vessel itself.

Crew complements of typical ships are as follows: galleys — 120 for a bireme, 200 for a trireme, of whom two-thirds would be oarsmen and the balance marines and sailors to handle the rigging and steering; sailing ships — in the Elizabethan period the numbers were about the same, the crews of Spanish ships during the Armada averaging just over 200, including soldiers, those of the generally smaller English vessels averaging 100 to 120, but by the time of the Napoleonic Wars larger ships of around 100 guns would have 8-900 sailors and marines, smaller vessels approximately 4-600 and frigates 150 to 300.

All firing, up until the introduction of fire-control radar in the Second World War, was by line of sight only and then at comparatively short ranges because, without stabilizing and compensation systems, the rolling and pitching of any vessel made its guns highly inaccurate. Thus, although a large first rate ship of the line of Nelson's time could have as many as 120 guns composed of 32, 24, 12 and 9 pdr guns and frigates thirty to forty 18, 12, 9 and 6 pdrs, the range at which they could begin firing would normally have been about the same — approximately a mile — but they would not become effective until well under half that. It is the number of guns carried, therefore, which is really more important than their size, especially since most battles were fought at virtually point-blank range once the fleets were physically engaged. For the purposes of simplicity I am also going to assume that warships with revolving turrets would normally fire salvoes at individual targets rather than engaging separate targets with each turret, so all firing is by 'broadside'. This is not, of course, strictly accurate, and vessels of the steam and modern eras may engage targets from head-on or astern at half effectiveness.)

In the age of sail we assign a 'fire factor' to each vessel depending on the number and calibre of guns carried in each side, and as explained earlier I have 'bracketed' these.

A first, second or third rate ship with three gun decks and over eighty guns has a fire factor of 50; a two-deck vessel with over forty guns a factor of 30; and smaller vessels a factor of 10. (You could easily make this listing more comprehensive if you wished — the above factors are based on vessels with 100, 60 and 30 guns ranging from 6 to 32 pdrs.) Each fire factor is for a broadside, ie, 30 points each side for a two-decker. All vessels have the same firing ranges: 1,000 yards (thirty inches) is extreme range, 500 yards (fifteen inches) is medium range and 200 yards (six inches) close range. At all except close range it would have been impossible to fire chain shot at an enemy's rigging, and all hits are therefore classed as on the target's hull; at close range you can select either chain or round shot, and this should be written into your orders for each turn. As in armoured warfare, the procedure is to declare your target before measuring the range. If you have loaded with chain shot and find the target is more than six inches away, the shot hits the hull not the rigging, with halved effectivess (ie, half of the damage points which would otherwise have been inflicted).

Each vessel starts a battle with a 'points value' equal to three times the total number of guns carried, so the three classes of ship given above would have starting values of 300, 180 and 90 points respectively. Damage inflicted is deducted gradually from these values until a ship surrenders ('strikes') or is sunk. The procedure is as follows: throw one average dice for every five fire points. Deduct -1 from each throw at extreme range, add +1 to each at close range. The total represents the number of damage points which must be deducted from the starting value of the target vessel. Add a further +1 at medium or close range if firing at the stern or bows of the enemy ship ('raking').

Damage points inflicted are taken off the hull only at extreme range, off hull and guns at medium range and off hull and guns *or* rigging (if loaded with chain shot) at close range, the total points value of each ship at the start being made up in three equal portions — thus a three-deck ship has 100 hull points, 100 gun points and 100 rigging points.

A ship's speed is reduced by one knot for every ten damage points inflicted on its hull and by one knot for every five damage points inflicted on its rigging. A ship which is halted in the water is considered to be sinking and its gunners will abandon their positions to take to the boats or cling to pieces of wreckage in the hope of being picked up after the battle.

At medium and short ranges damage scores are divided equally between hull and guns. Ten damage points against the guns means the player commanding that ship throws one less dice next time he fires.

Vessels suffering twenty damage points to the rigging must heave-to for three turns to cut the fallen masts and sails away.

When a vessel has suffered damage points totalling a third of its original value, a morale test must be taken to see whether it will surrender or attempt to flee, and at the end of every turn thereafter. Throw one average dice. With a four or five it will continue the battle; with a three it will attempt to escape; with a two it will surrender and be taken as a prize by the winning fleet at the end of the battle. Add +1 to the dice score if the vessel is the fleet's flagship. If the flagship surrenders, so will the remainder of the admiral's ships.

Boarding

If two vessels are brought alongside each other, a mêlée will be fought between the crews using the rules as for land warfare, and taking a morale test at the end of each turn. All sailors and marines are classed as fighting with 'other infantry weapon' and are 'disordered'. Marines class as regulars, sailors as irregulars. After each round of mêlée take a morale test as in the rules for land warfare. A three-deck ship counts as 'uphill' from a one or two-decker, and so on. A morale score of four or less means the ship surrenders, otherwise it fights on.

Ramming (galleys and ironclads)

In order to successfully ram an enemy vessel, it must be hit in the side (not bow or stern) head-on and at a speed of at least four knots. If successful, the enemy ship will begin to sink, this taking three turns in a light wind, two turns in a moderate wind and one turn in a strong wind. Its crew therefore has from one to three turns to attempt to take over your own ship by means of a mêlée, but will surrender the moment their own ship sinks. If you do not succeed in ramming head-on, because you have miscalculated your approach or the enemy evades, the ram will break half the oars on that side of the enemy ship, reducing its speed by half. Your own galley may sail on or engage.

Dreadnoughts and later

Vessels are again assigned 'points values' at the start of the game and lose fighting effectiveness as these are eroded. To all intents and purposes there is no 'rigging' to consider and no modern equivalent of chain shot, but there are boiler rooms and engines instead. Ships' points are calculated as follows and here it is again necessary to note that the figures given in the rules are for representative vessels of each class and that there were wide variations between individual ships within each broad class, so you may wish to refine these figures. Points are based upon armament carried but damage is inflicted in four proportions against hull, propulsion gear (engines and screws), gun turrets and superstructure (command systems, etc). Starting points are as follows: battleships (both World Wars) — 1,000; battlecruisers — 900; heavy cruisers — 800; light cruisers — 600; destroyers — 400; frigates — 300; submarines — 100; fleet carriers — 1,000; escort carriers — 500.

Since a carrier's main armament was its aircraft, half of all damage points to hull, armament or superstructure are taken off the flight deck, which effectively counts as a fifth target zone, and once a flight deck has received thirty per cent of its value damaged the carrier is incapable of flying aircraft on or off. (The points value of a flight deck is 200 for a fleet carrier and 100 for an escort.)

'Broadside' or fire points values are: 300 for a battleship, 270 for a battlecruiser, 240 for a heavy cruiser, 180 for a light cruiser, 120 for a destroyer and 100 for a frigate. Fleet and escort carriers count as destroyers and frigates respectively. Submarines use torpedoes only, see below, but destroyers can use torpedoes as well as guns. The dice-throwing procedure is as before but you throw one average dice for every *thirty* fire points and then multiply the totals by three. This is just a time-saving device to avoid having to throw an inordinate number of dice. Damage points inflicted are taken off the four vulnerable areas in equal proportions. Heavy cruisers and above (including fleet carriers) are assumed to have four screws, and lose a quarter of their speed for every 25 per cent damage on their propulsion gear; all other vessels are assumed to have only two screws, and lose half their speed for every 50 per cent damage on their propulsion gear. All ships have their speed reduced by half after suffering 25 per cent damage on their hulls, and sink after receiving 50 per cent.

The ranges of weapons for this later part of the naval period are such that few wargames tables are large enough to accommodate them realistically, and the effectiveness of heavy armour-piercing high explosive shells is less reduced by distance than in the era of solid shot, so a compromise solution is needed to make games playable. However, the secondary armament on a heavy cruiser or above was often superior to that of a light cruiser or below, but could only be brought into play once in range, so we again have three range brackets but with further refinements. These are mainly introduced to give different ships advantages and disadvantages when in reality any of their guns could fire over the whole area of a wargame table with almost equal effectiveness, so ranges have to be artificially 'compressed'.

Weapon ranges in feet (divide by 100 for inches)

	Maximum	**Medium**	**Short**
Battleships	Any	10,000	5,000
Battlecruisers	Any	9,000	4,500
Heavy cruisers	10,000	7,000	3,500
Light cruisers	8,000	5,000	2,500
Destroyers	6,000	4,000	2,000
Frigates	4,000	2,500	1,250
Torpedoes	4,000	2,000	1,000

(**NB** Carriers count as destroyers or frigates as stated above.)

At 5,000 yards and less, heavy cruisers and above add +1 to each dice throw. At 2,500 yards and less, light cruisers add +1 and heavy cruisers and above +2.

Torpedoes may be fired by submarines or destroyers. Submarines are considered to have two tubes, destroyers four. Throw one ordinary dice per tube. At maximum range a six is required for a hit, at medium a four to a six and at minimum a two to a six. Each torpedo hit scores 200 damage points against the hull of the target only.

Damage points on guns reduce the number of dice the target vessel can throw on the next turn by one dice for every thirty points of damage inflicted. Hits on the superstructure reduce the number of dice by one for every fifty points inflicted.

Submarines

Since they would normally operate underwater, introducing submarines to a battle causes problems because you cannot place a model boat on the table without your opponent seeing it, which would not normally be the case in real life. The way around this is to plot your submarines' movements on a gridded plan of the wargames table, on which are also marked the positions of all other ships at the beginning of each turn. When torpedoes are fired, it is assumed that the general direction from which they have come is known to the other player because of where a target is hit, and this must be revealed. You should announce that a submarine is firing at the beginning of your turn and then throw a dice, but you do not reveal direction until you have scored a hit, and then not the range. Enemy destroyers will search along the approximate line of the torpedo(es) track, using sonar to try to detect it. This was very hit and miss, so throw one normal dice for each destroyer or frigate at the beginning of its turn. A six means it has found an echo and you should then indicate the approximate area of the table where the submarine is hiding, within a twelve-inch circle. On the next turn, if the submarine has moved outside this circle, each destroyer again needs a six to remain in contact, or a five or six if the submarine is still within the circle, in which case you must pinpoint its position within six inches. On the third turn the destroyer needs a five or six to stay in contact if the submarine has moved outside the six-inch circle, but only a three to a six if it is still inside, when you must reveal its location to within three inches. Once inside this three-inch circle, the destroyer can drop depth charges, six per turn, throwing one ordinary dice for each. Two consecutive sixes are needed for a 'kill', one six to damage the submarine so that it cannot manoeuvre but must either surface or hope the destroyer goes away. A damaged submarine may fire no more torpedoes.

As I am sure you will appreciate, all of the foregoing systems can be refined, added to and made much more realistic, and one of the most enjoyable facts about wargaming is that there is no single hard and fast set of rules, such as in Chess, to which everybody plays. Each wargamer has his or her own ideas and your own knowledge of a particular subject can lead to suggestions which can improve any rules. The above is a prime example. With a home computer, you could 'hide' your submarine, alter its depth and course without your opponent's knowledge, while he could carry out proper search pattern sweeps and set his depth charges to explode at different depths. This is warfare, of course, in three instead of the previous two dimensions, and becomes even more complicated when one comes on to dogfights during the Battle of Britain or similar...

Meanwhile, those readers wishing to explore the naval possibilities more thoroughly should read Paul Hague's book *Sea Battles in Miniature*, also published by PSL but now unfortunately out of print. A copy can, however, be ordered through your local library.

AERIAL WARFARE

While the First World War is generally unpopular amongst both 'land' and naval wargamers, exactly the reverse is true when one comes to aerial wargaming, and almost any wargames exhibition will feature at least one table with gaily painted Albatrosses, Pfalzs and Fokkers mixing it with SE 5s, Camels and the occasional RE 8 hoping no-one is going to attack him. There are various reasons for this but one of the principal factors is the supposed 'glamour' of the period with its chivalrous 'knights of the air', a myth perpetuated for generations of schoolboys by Biggles and similar stories. In reality there was precious little chivalry, the main aim being to shoot the other blighter down with as little risk as possible to oneself. 'Shooting a man in the back' assumed a different meaning as the approach from the rear, hopefully with a height advantage and out of the sun or clouds so that the enemy pilot did not spot you until too late, became the favourite tactic to adopt.

As far as wargamers are concerned, however, one major attraction of the First World War is the comparatively low speed of the aircraft, since even late war machines such as the Spad S XIII, Albatros D Va or Nieuport 17 were incapable of speeds much in excess of 100 mph and most dogfights were fought at less than this. Compare these sorts of figures with a World War 2 Typhoon or Tempest bombing along in excess of 400 mph, or an Eagle, Tomcat, MiG-25 or Tornado at Mach 2+ in a modern air combat environment, and the advantages of the First World War become even more obvious. If using 1:300 scale model aircraft, for example, as in armoured warfare, a true air scale nevertheless makes a mile equal to 17.6 feet and an aircraft flying at 100 mph would traverse this in 36 seconds, so it is obvious that some wargaming compromises are again necessary.

The First World War saw the transformation of a tentative and comparatively untried military device into a scientific and efficient weapon. In 1914 the average military aircraft could not exceed 70-80 mph and maximum height (the 'service ceiling') was less than 10,000 feet, while armament consisted of little more than a pistol or rifle fired single-handedly by the pilot since the principal role of the aircraft was seen as that of reconnaissance and forward spotting for the artillery. Some two-seater aircraft were fitted with a machine-gun in the rear cockpit for self-defence by the observer, but this was rapidly found inadequate and aircraft were gradually fitted with one, and then two, forward-firing machine-guns. This in itself posed problems, because they had to be mounted on the top wing of either biplanes or triplanes, making aiming by the pilot in the cockpit below extremely difficult and reducing his control over the aircraft. The answer was to place the machine-guns in the forward fuselage directly in front of the pilot, but then the propeller got in the way unless the aircraft was of a 'pusher' design like the Vickers Gunbus, with the propeller behind the pilot. However, such designs were slower and less manoeuvrable than those with 'puller' propellers, and an early innovation was the fitting of deflector plates to the rear of each propeller blade so that, when a machine-gun bullet struck it, it did not damage the blade. This proved hazardous to the pilot as bullets were often deflected directly backwards! Eventually, interrupter mechanisms came into use, as on the Sopwith Pup of 1917, which stopped the guns firing when a blade was directly in front of them, and eventually hydraulic synchronizing gear as on the Camel and SE 5a, for example.

The first true fighter with a fixed forward-firing machine-gun to emerge was the German Fokker Monoplane in 1915 and the Allied British, French and Belgian machines found themselves at an immediate and distinct disadvantage, but aerial battles were rare until the beginning of the battle of the Somme in July 1916. Flying low over the trenches, the German 'flying circuses' pioneered by Boelcke began by running rings around the Allied aircraft, but new designs and tactics came forward swiftly under the brutal impetus of war and a see-saw struggle for aerial supremacy emerged. The scale of this can be appreciated when it is realized that, at the beginning of the war the fledgling Royal Flying Corps possessed 63 aeroplanes, while at its end the Royal Air Force had 3,300 in the front line. Bombers had also been developed alongside fighters, and whereas the earliest types were just Scouts whose pilots could drop hand grenades over their cockpit sides, by 1918 heavy twin-engined machines carrying bombs weighing hundreds of pounds and capable of reaching Berlin were being built.

During the inter-war years there was a great deal of progress in aircraft design although the biplane was still favoured by most air forces because of its manoeuvrability and because many people felt that monoplanes were 'unsafe'. Until practically the beginning of the Second World War, for example, the Royal Air Force was almost exclusively biplane-equipped, the Hawker Hurricane not making its first flight until 1935, nor the Spitfire until 1936, both types entering service in 1937 alongside the Bristol Blenheim, a twin-engined bomber which startled observers of the time by being faster in level flight than any existing biplane fighter. In Germany the Messerschmitt Bf 109 had also first flown in 1935 and the first production Bf 109B-0s joined the Luftwaffe in 1937; these developments were soon matched by other countries.

The most significant advances in military aircraft design in the inter-war years lay in engine manufacture as they became lighter and more powerful, permitting speeds in excess of 300 mph. There were constant arguments between the proponents of water-cooled in-line engines, which could be made slimmer and therefore enable aircraft to be more streamlined, but which were more vulnerable to enemy fire; and air-cooled radial engines which were more robust and required less maintenance. It is interesting that the RAF and Luftwaffe both entered the war with in-line designs in the Hurricane, Spitfire and Bf 109, but that the most successful fighters to emerge during the war were both radial-engined, the Focke-Wulf Fw 190 and the Tempest.

Hand-in-hand with engine developments came improvements in armament. Biplanes, because of their thin wings, had to have their machine-guns mounted in the fuselage, but thicker-winged monoplanes could have them in their wings so the number could be increased, first to four and then to eight. Alternatively, weapons of heavier calibre could be fitted, such as 20 mm cannon, and again there were exponents of both types, one school of thought reasoning that a fighter pilot stood more chance of hitting his target if he had multiple guns spraying a veritable hail of fire, the other arguing that heavier guns had greater stopping power. Eventually it was the latter which emerged as the most successful formula and by the end of the war almost all fighters were cannon-armed, including the first operational jets such as the Gloster Meteor and Messerschmitt Me 262.

Bombers also evolved, but the design criteria were different. The principal function of a bomber is obviously to drop the heaviest tonnage of high explosive possible on to an enemy target — but the aircraft has to reach its target first and then, hopefully, return, so other factors also come into play, including speed, height and defensive armament to protect the bomber from enemy fighters. One of the gravest mistakes Germany made in the Second World War was to devote insufficient attention towards developing a strategic long-range bomber comparable to the American B-17 or the RAF's Lancaster and, however devastating their attacks seemed to the recipients of their attentions in 1940 and '41, the German Heinkel He 111s, Dornier Do 17s and '217s and others were incapable of maintaining a heavy offensive and had to be protected by fighters of limited endurance in order to reach their targets at all.

Strangely, two of the most effective fighters of the war started life as bombers, the Junkers Ju 88 and the de Havilland Mosquito, both reliant upon high speed for their protection. All bombers, other than fighter-bombers able to attack a target and then defend themselves were multiple-engined, four propellers being normal on the larger, long-range machines. (An odd exception was the German He 177, designed as a strategic bomber but incapable of carrying a heavy enough payload, which had four engines driving two propellers.) Such bombers usually bristled with gun turrets fore and aft as well as ventral, dorsal and in their sides, but were still vulnerable to enemy fighters both in daylight and, after the introduction of airborne interception radar, at night. They therefore relied upon long-range fighter escorts such as the P-38 Lightning, P-47 Thunderbolt or P-51 Mustang to defend them. There were even plans, late in the war, to equip each bomber squadron flying B-29 Superfortresses with an aircraft modified to carry a fighter underneath each wing! The end of the war caused the demise of such ideas, but that they could have been successful is shown through post-war American high-speed and altitude research aircraft which were normally carried aloft by a 'mother ship' before launch.

In the post-war world, the jet engine has, of course, been the single most significant factor, producing aircraft capable of ever-faster speeds and endurance. However, increasingly sophisticated radar and computer technology, particularly in the field of micro-miniaturization, has produced other changes, not least being the virtual disappearance of guns from all except ground-attack aircraft. Instead, today aircraft are armed with a wide variety of guided missiles which track their targets through infra-red, radar and television sensors. To begin with, such missiles had to be tracked to their targets by the pilot or a second crew member aboard the firing aircraft, but today most are of the 'fire and forget' variety, instrumented with sophisticated computer guidance systems which will take them unerringly to a chosen target on their own.

Air battles in miniature

Aerial warfare is by far the most difficult to recreate realistically not just because there is an extra dimension to cope with — the vertical — but because each different aircraft type has its own individual characteristics such as rate of climb, maximum ceiling, stall speed, turning radius, rate of roll and even different possible speeds at different altitudes. For these reasons this section cannot be comprehensive and can only deal with aerial dogfights between two pairs of representative fighters from each of the World Wars. Readers seeking more information should acquire a copy of Mike Spick's definitive book on aerial wargaming, *Air Battles in Miniature*, also published by PSL. Like its companion volumes on naval and armoured warfare, this is unfortunately out of print but your local library should be able to help. It covers not just fighter-versus-fighter combat but also air-to-ground and ground-to-air as well as aerial campaigns and at the time of writing is certainly the best work on the subject.

The height dimension can be catered for in various ways, none of them wholly satisfactory, such as fixing model aircraft by means of sliding clamps to rods marked out in a scale of feet, but this tends to get extremely confusing and in the end Mike Spick hit on the revolutionary concept of dispensing with the *horizontal* 'sideways' dimension entirely, since his analysis of fighter combat showed that the crucial factors were height and speed coupled with rate of roll and stall speed rather than an aircraft's ability to bank or turn on the level. The wargames table thus has one long edge representing the ground and the opposite edge the maximum altitude of the model aircraft being used. Ordinary model aircraft, in particular the popular 1:300 scale ones, can be used with this system, although it looks odd having all your aeroplanes constantly flying on their sides with one wingtip pointing at the ground. Mike Spick therefore decided to cut his models in half, vertically, each half thereby representing an aircraft flying from left to right or right to left.

The most suitable ground and time scales for aerial wargaming in this way Mike found after considerable trial and error to be one millimetre to four yards and four seconds to a move or turn, giving the average eight by six foot wargames figure a baseline of 5½ miles and a ceiling of just under 22,000 feet, quite acceptable for

Opposite and overleaf *Aerial combat the Mike Spick way, using 1:144 scale Spitfires and Messerschmitt Bf 109s cut in half as described in the text. The use of the two alternate halves as aircraft change direction can clearly be seen as the '109s 'bounce' the Spitfires. Fortunately the wing second is alert and at the end of the manoeuvre one '109 is well placed in the wing second's sights while the wing leader has managed to evade his pursuer* (models by Mike Spick, photography Brian Monaghan).

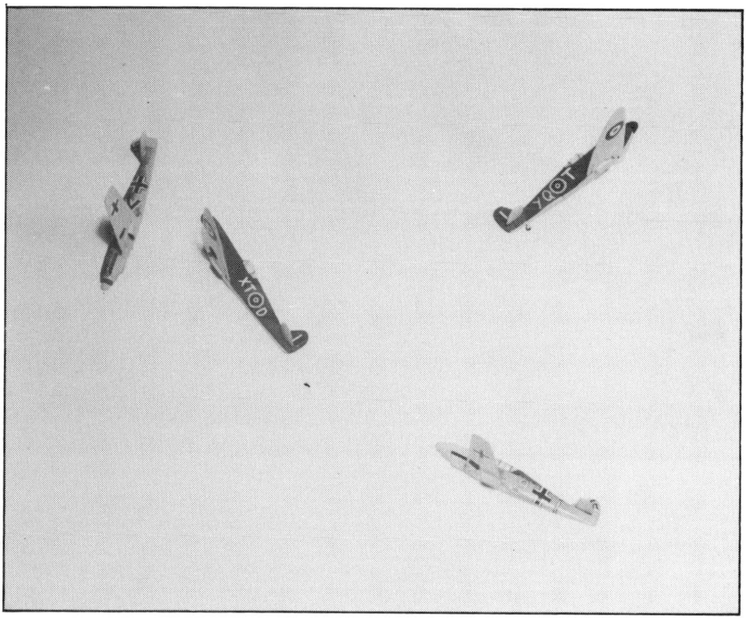

most dogfights. From this it can be seen that a model aircraft travelling at a scale 100 mph covers 50 yards a second or 200 yards in a move which, at one millimetre to four yards gives a game move of 50 mm. In other words, halve the actual speed in mph and you instantly have the tabletop speed in millimetres.

Now, although we have not got a horizontal dimension for aircraft to turn in on the table, we must allow for it in another way or they will simply fly in from one end of the table and out the other, unless they loop, which is not always desirable. Like warships, covered earlier, we need a form of turning circle to denote the amount of space in the sky an aircraft needs in which to turn around. Although other factors do come into play, such as angle of bank and the amount of 'g' force the aircraft (and its pilot!) can stand, for practical purposes turn rates can be said to be directly proportional to speed and that an aircraft travelling at 100 mph will need 100 mm in which to turn through 180° so it is flying back in the opposite direction to that in which it started, one flying at 400 mph will require 400 mm, and so on. No 'turning circle' is thus required, the aircraft simply reverses direction (ie, the model is replaced by its other half) when it has covered the distance over the ground dictated by its speed. In his book, Mike actually gives much more accurate figures than this depending upon an aircraft's wing loading, but the above is good enough to give you the 'feel' of the system.

As with warships, again, aircraft did not bomb around the sky at maximum speed all the time so we need to devise a system for acceleration and deceleration, and this is simply taken care of by saying that an aircraft in level flight can accelerate or decelerate by opening or closing the throttle at ten per cent of its maximum speed per turn — thus an aircraft with a top speed of 300 mph could gain or lose 30 mph (15 mm) per turn up to its maximum or down to its stall speed. (A stall occurs when the airflow across the wings is insufficient to keep the aircraft flying, and it will enter a spin which can only be corrected by putting the nose down until flying speed

1:300 scale models can also be used but 'fly' on their sides (models by Mike Spick, photography Brian Monaghan).

has been regained.) Since stalling speeds and rates of climb and dive were different for each individual aircraft type, I have just given these in the following table for my chosen four fighters, but figures for other aircraft can easily be found in the many reference books around. As a rule of thumb, acceleration in a climb is one-third that of level flight, so the preceding aircraft with a maximum speed of 300 mph could gain 10 mm in height while gaining 10 mph in forward speed. However, aircraft with good rates of climb do not always have good rates of dive, because the latter is more directly related to the aircraft's weight, so a light aircraft which could overhaul a heavier one in the climb would lose it in the dive, and dogfight tactics have to be adapted to suit an individual aircraft's characteristics.

Having learned how to manoeuvre our aircraft, we must now bring them into battle, and the first consideration is observation. Most aircraft had blind spots to the rear and below in particular and

we must cater for the pilot of one aircraft not seeing an enemy until cannon shells start streaking past his cockpit. For this we use the conventional clock code converted to the vertical plane, so 12 o'clock is always vertically above the pilot's head, regardless of the aircraft's attitude in the sky (ie, climbing, diving or even inverted), 3 o'clock is directly ahead of the aircraft's nose, 6 o'clock is vertically beneath it and 9 o'clock is dead astern — 'dead' often being the operative word! Using an average dice, throw a three, four or five to detect an enemy aircraft flying ahead and/or above you in the 12 o'clock to 3 o'clock arc; a four or a five if it is ahead and/or below you in the 3 to 6 o'clock arc; a four if it is astern and/or below you in the 6 to 9 o'clock arc; and a five if it is astern and/or above you in the 9 to 12 o'clock arc. A two always means you did not spot the enemy. These dice throws are repeated at the beginning of each turn until the enemy is spotted or he spots you and attacks. If an aircraft dives into cloud you do not know where it is going to emerge with any precision but you do know he's there somewhere, so add +1 to your dice throw each time when searching for him again.

Clouds are represented on the wargames table by pieces of ceiling tile cut to suitable irregular shapes. An aircraft in cloud may manoeuvre as normal and does not need to emerge from the other side travelling in the same direction as when it entered the cloud.

Having spotted your enemy, you now have to shoot him down, which isn't as easy as it sounds unless you have him in your sights from astern and he hasn't noticed you are there. Deflection shooting, with a target moving at an angle across your line of fire, is always tricky since the target may be in your sights for only a fraction of a second, and relative speeds are thus also important. Using a protractor, measure the angle at which the target aircraft is crossing your nose. At 60° to 90° you need to throw a five on an average dice to score a hit; at 30° to 60° a four or a five; and at 0° to 30° a three, four or five. All twos are misses. Deduct -1 from your score for every 50 mph faster than your own aircraft the target is

Aircraft	**Stall speed**	**Cruise speed**	**Top speed**
Sopwith Camel	25mph	105mph	120mp
Fokker D VII	25mph	100mph	115mp
Spitfire V	70mph	325mph	375mp
Messerschmitt Bf 109F	75mph	340mph	390mp

moving (ie, if your own aircraft is travelling at 200 mph and the target at 250, deduct -1).

Hits, of course, are not always immediately decisive, many aircraft being able to sustain an enormous amount of damage and still make it back to their airfield safely. During the Second World War the American Navy conducted a survey which showed that forty per cent of hits on single-engined aircraft were to the fuselage, twenty per cent on the pilot and/or controls, ten per cent on the engine or propeller, twenty per cent on the oil, fuel or hydraulic systems and ten per cent elsewhere. Although not strictly in proportion, having achieved a hit throw another two average dice to give the following: a pair of threes or fours gives a fuselage hit; a three and a four a pilot hit; a three and a five an engine hit; a four and a five a hit to one of the liquid systems; and a pair of fives a hit on the control system. Twos are discounted — if you get one, throw that dice again.

Range and the calibre of your guns will determine the amount of damage caused. Maximum effective range for cannon is 1,000 yards (250 mm), for machine-guns 500 yards (125 mm). At these ranges each cannon has a 'fire factor' of two, each machine-gun a factor of one (thus a four-cannon fighter will have eight factors at maximum range, comparable to that of an eight machine-gun fighter at half that same range). Add +1 for every twenty per cent closer than maximum the target is (thus +1 every 200 yards (50 mm) for cannon, +1 every 100 yards (25 mm) for machine-guns). The total represents the damage points inflicted on the target, maximums being 24 for a four-cannon fighter at under 200 yards or the same for an eight machine-gun fighter at under 100 yards. Each part of an aircraft starts the game with 20 defensive points, and when 20 are knocked off any of the systems hit results are as follows: on the fuselage — if the aircraft can escape its attacker it will head straight for the nearest airfield to land; on the engine — the engine is dead and the pilot will bale out; on the pilot — he is killed

Service ceiling	Rate of climb	Rate of dive	Armament
24,000ft	880fpm	800fpm	2 x MG
23,000ft	1,025fpm	900fpm	2 x MG
37,000ft	2,700fpm	3,000fmp	4 x Ca
39,000ft	3,450fpm	3,000fpm	4 x Ca

and the aircraft dives into the ground; on the fuel systems, etc, — aircraft catches fire and may explode; throw an extra dice — two or three, aircraft explodes, pilot killed; four or five, pilot bales out safely; on the control systems — aircraft is unable to manoeuvre and must head for home in a straight line. (Baling out, incidentally, is an option only open to Second World War pilots, for in the First parachutes were not normally available until right towards the end of the war, so a pilot whose aircraft was killed was also dead.)

Playing factors for my chosen aircraft are as shown in the chart on the previous spread.

There are some interesting figures here, as you can see, which make aerial wargaming so fascinating. The First World War Camel and D VII have very similar speeds and service ceilings, and effectively the same armament, but the D VII was better in the climb because of a more efficient power-to-weight ratio and better in a dive because it was considerably heavier. In the Second World War the Bf 109F appears to outclass the Spitfire V in virtually every respect, yet in battle they were evenly matched, proving that a great deal lay in the aggressiveness and skill of the pilot — or, in this context, you, the wargamer.

Rates of climb and dive should simply be divided by 100 and translated into millimetres per turn on the table. MG = machine-gun, Ca = cannon.

My apologies and thanks to Mike Spick, first for the way in which I have simplified them and secondly for allowing me to 'borrow' his rules. I hope you now read his book, and his also excellent *Fighter Pilot Tactics* which explains the actual techniques of daylight air combat from World War 1 to the present day.

FANTASY AND OTHER GAMES

It is an odd truism that wargaming seems to attract people who also read a lot of science fiction or fantasy-type novels, perhaps because both are essentially escapist in nature. Philosophy aside, one of the greatest phenomena within wargaming over the last ten years or so has been the growth of the fantasy role-playing game which masquerades under various guises, the most popular still being the 'Dungeons and Dragons'™ system originated in America. Such games have two basic origins, one being science-fantasy novels such as those of Michael Moorcock and other writers who have created esoteric worlds populated by heroes and heroines wielding swords and magical weapons with equal dexterity; the other being the earlier skirmish-type wargame pioneered in Britain originally for such scenarios as an 'OK Corral' style Western gunfight.

The idea behind the original skirmish games was that each player could have one, or sometimes more, model soldiers on the table, with whom to identify personally. Each figure had its own 'characteristics' decided by initial dice throws and later through 'experience points' gained after surviving gunfights. These covered such factors as speed in drawing guns, accuracy, physical dexterity in such matters as jumping on and off a horse and so on, just like a stunt man in the films, and the games were geared to individual interreaction between the figures on the table — which usually featured elaborately constructed Western townships. I helped take part in an exhibition game of the Alamo using such rules once, but only after the organizers agreed to let me have Davy Crockett as my character!

Like many wargamers, I read a lot of science-fiction, but to begin with I was not attracted by fantasy wargaming. Wargaming, as far as I was concerned, involved accurate historical research and an attempt to recreate the real warfare of a chosen period as accurately as possible using the chronicles and analyses of the time. Adventures in dismal dungeons populated by denizens as unlikely as they were malignant seemed a far cry from this. Then some friends started a game one evening and, with nothing else to do, I joined in and took the character of a warrior/thief based loosely upon Fafhrd and the Grey Mouser in the superb Fritz Leiber stories. I rapidly found myself entering into the part of the character — not a nice one, rather than finding 'loot' for himself he developed a propensity for letting others do the dirty work then stealing it for

A wizard roasts a haunch of some mythical beast during a fantasy role-playing game (figures by Phil Barker, photography Brian Monaghan).

himself, but fun to play. This became my undoing in a later, full-scale campaign game when I played the role of a werewolf... but that's another story.

In a fantasy game, each player assumes the persona of a character on the table, the usual roles being wizard or witch, thief, cleric or warrior, with various subtle sub-divisions. The game, unless it is a commercial package, will be designed by one of the players who is usually known as the dungeon-master, game-master or lore-master, and who acts as umpire as well as croupier dealing out nasty surprises. Settings vary from fantasy worlds — often involving castles and labyrinths — of the designer's own imagination, to well-sculpted worlds based upon the established writings of Moorcock and others or upon historical models such as the dark ages, with the main difference that magic actually does work! The games usually take the form of a quest of some description, for the Holy Grail, or some other treasure such as a maiden imprisoned by a dragon, and they involve both hand-to-hand combats, magical duels and the solving of logic puzzles such as riddles as well as the avoidance of concealed traps and dangers such as fluorescent green creeping fungus... Fantasy role-playing games of this type are not really wargaming in the traditional sense, but they are immensely popular so you should give one a try at some point.

In the same class as far as I am concerned fall science-fiction games of the 'Star Wars' variety, but here there is more similarity with traditional games in that spacecraft and weapons systems can be allocated ranges, speeds and offensive and defensive capabilities like an aircraft or battleship and a good game requires a reasonable knowledge of astronomical phenomena, ballistics and similar subjects, so genuine research is needed. Again, though, novels such as Dr E.E. Smith's classic 'Lensman' series provide the main background setting and motivation, and you need to be well versed in them in order to anticipate your opponent's moves.

Home computers have come into their own as wargames tools in the last few years, and can either be programmed to keep track of supplies, casualties and the like during a 'conventional' wargames campaign, as well as computing results on the table to save endless counting up of factors and dice-throwing; or they can be programmed to display games on a television screen without any need for model soldiers or other paraphernalia. There are several commercial games of this type available, most of them featuring platoon-type skirmishes set during World War 2 or science-fiction scenarios, and these are regularly advertised in both computer and wargaming magazines.

BIBLIOGRAPHY

For ancient and mediaeval warfare, the best books are the extensive series published by the Wargames Research Group, 35 Long Street, Devizes, Wilts, SN10 1NT, a full list of which is available on request. At the time of writing these cover the armies, weapons and tactics of practically every nation from the earliest times up to the 15th century. WRG also publish extensive and widely used sets of playing rules, and army lists to enable newcomers to the hobby to build up balanced forces.

Most of the 'classic' works on wargaming are sadly out of print, but still obtainable through libraries (if your local branch does not have one in stock, they will order it for you at a modest charge). These include: *War Games, War Games Campaigns* and *Advanced War Games*, all by Donald Featherstone and published by Stanley Paul; Charles Grant's *Charge!*; the five volumes of *Tank Battles in Miniature* by Donald Featherstone and myself, all published by Patrick Stephens; *Air Battles in Miniature* by Mike Spick and *Sea Battles in Miniature* by Paul Hague, also published by PSL; *Skirmish Wargaming* by Donald Featherstone, and *Napoleon's Campaigns in Miniature* by myself, again published by PSL (the latter now being available in paperback).

Individual works on the warfare of different periods are too numerous to attempt to list but a catalogue of all significant books available, including many antique and out-of-print titles, can be obtained from The History Bookshop, 2 The Broadway, Friern Barnet, London, N11 3DU, while others are regularly advertised in magazines such as *Military Modelling* and *Wargames*, both published by Argus Specialist Publications in Hemel Hempstead.

INDEX